Charles Seale-Hayne Library
University of Plymouth
(01752) 588 588

HPLC
A Practical Guide

RSC Chromatography Monographs

Series Editor: Roger M. Smith, *Loughborough University, Loughborough, UK*

Advisory Panel: J.C. Berridge, *Sandwich, UK*; G.B. Cox, *Illkirch, France*; I.S. Lurie, *Virginia, USA*; P.J. Schoenmaker, *Amsterdam, The Netherlands*; C.F. Simpson, *London, UK*; G.G. Wallace, *Wollongong, Australia.*

This series is designed for the individual practising chromatographer, providing guidance and advice on a wide range of chromatographic techniques with the emphasis on important practical aspects of the subject.

Packed Column SFC
by T.A. Berger, *Berger Instruments, Newark, Delaware, USA*

Chromatographic Integration Methods, Second Edition
by Norman Dyson, *Dyson Instruments Ltd, UK*

Separation of Fullerenes by Liquid Chromatography
edited by K. Jinno, *Toyohashi University of Technology, Japan*

HPLC: A Practical Guide
by T. Hanai, *Health Research Foundation, Kyoto, Japan*

How to obtain future titles on publication

A standing order plan is available for this series. A standing order will bring delivery of each new volume immediately upon publication. For further information, please write to:

The Royal Society of Chemistry, Turpin Distribution Services Ltd., Blackhorse Road, Letchworth, Hertfordshire SG6 1HN, UK.
Telephone: Letchworth +44(0) 1462 672555: Fax: +(44)0 1462 480947

RSC
CHROMATOGRAPHY
MONOGRAPHS

HPLC
A Practical Guide

T. Hanai
Health Research Foundation, Kyoto, Japan

ROYAL SOCIETY OF CHEMISTRY

ISBN 0-85404-515-5

A catalogue record for this book is available from the British Library

Published by The Royal Society of Chemistry,
Thomas Graham House, Science Park, Milton Road
Cambridge CB4 0WF, UK

For further information see our web site at www.rsc.org

Typeset by Paston PrePress Ltd, Beccles, Suffolk
Printed by Bookcraft (Bath) Ltd

Preface

This book answers two basic questions regarding high-performance liquid chromatography; it focuses on how compounds can be separated and on why particular compounds are separated by liquid chromatographic methods. It consists of six chapters: Basic Concepts of HPLC; Instrumentation: Preparation, Testing and Selectivity of Stationary Phase Materials; Selection of the Eluent; Separation Based on an Improved Column Efficiency; Influence of Physical Chemistry on Separations in Liquid Chromatography. The focus is on the basic considerations in liquid chromatography rather than the applications but there are also sections on trouble-shooting. The book is concerned mainly with the selection of a packing material and the preparation of the eluent. Theoretical optimization is demonstrated using the chromatography of simple chemicals, to aid the understanding of liquid chromatography by graduate students. The examples given go beyond what can be found in general textbooks on liquid chromatography. The separations can be easily understood from the differences in properties of familiar compounds known to undergraduate students.

The chromatographic separations described are based on solubility using molecular properties. The selection of chromatographic mode and column is analysed using solubility parameters. The separation factor α is described in detail by physical and chemical parameters. Discussion of instrumentation focuses on high-efficiency operation and is concerned with degassing, the time constant, flow cell design, connectors and trouble-shooting. Packing materials and their synthesis and surface modification, liquid chromatographic analysis of surface activity and the evaluation of packed columns, including the measurement of void volume, are also considered.

I thank Professor Roger Smith, of Loughborough University, who reviewed the original manuscript and offered many helpful suggestions. I also deeply appreciate the support of The Royal Society of Chemistry for this publication.

Toshihiko Hanai, Yokohama

Contents

CHAPTER 1

Basic Concepts of High-performance Liquid Chromatography

The two basic questions in high-performance liquid chromatography focus on (a) how particular compounds can be separated, and (b) why particular compounds were separated by the liquid chromatographic method used. The answers can be obtained by the consideration of some simple representative chromatograms of the separation of well-known compounds. Such separations can be easily understood according to common principles of physics and chemistry.

A separation is described by the following equation, which indicates the degree of resolution between two peaks in a chromatogram, R_s. A complete separation requires $R_s > 1.2$ units.

$$R_s = \frac{1}{4}\left(\frac{\alpha - 1}{\alpha}\right)\sqrt{N}$$

The resolution can be improved by increasing the column plate number, N, and/or the separation factor, α (α = the ratio of the retention factors of the two compounds). N is the physical parameter and α is the chemical parameter for the separation. Higher N and α values give a better separation.

The physical and chemical aspects of liquid chromatography, in addition to mechanical aspects, are briefly described in this chapter. Theoretical approaches are explained in detail in later chapters. The effect of stationary phase materials on the chemical selectivity is described in Chapter 3, and the influence of the eluent components is covered in Chapter 4. The plate number theory is discussed in Chapter 5. Quantitative optimization is explained in Chapter 6.

1 Physical Parameters for High-speed Separations

It was thought that high-speed separations would be achieved by the development of a physically stable pumping system and highly sensitive detectors;

however, the main contribution to high-speed separation is made by small-size stationary phase materials. A shorter separation time with complete resolution cannot be achieved simply by increasing the flow rate or by using a small column. The theoretical plate number of a small column must be the same as that of a larger column to obtain the same separation.

For example, the separation of a mixture of benzene, acetophenone, toluene, and naphthalene has been completed within 5.5 min using a 15 cm long, 4.6 mm i.d. column, packed with 10 μm porous octadecyl-bonded silica gel, whose theoretical plate number was 38 000 m^{-1}, as shown in Figure 1.1A. Increasing the flow rate 4-fold reduced the separation time to 1.5 min, because this mixture was well separated (Figure 1.1B). The same mixture was separated within 4.5 min using a 10 cm long, 4.6 mm i.d. column packed with 3 μm octadecyl-bonded porous silica gel with a theoretical plate number of 117 000 m^{-1} (Figure 1.1C). Doubling the flow rate resulted in completion of the separation within 2 min, as shown (Figure 1.1D).

Comparison of these four chromatograms suggests that a fast separation can be performed either using a longer column with 10 μm stationary phase material with a high flow rate of the eluent, to give high resolution, or by a smaller

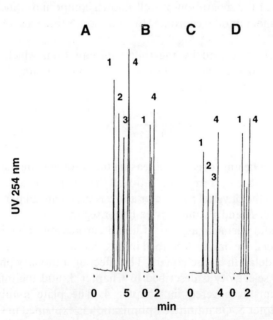

Figure 1.1   *Effect of particle size and flow rate on high-speed separations.* A *and* B *column*: 10 μm C$_{18}$-bonded silica gel, 15 cm × 4.6 mm i.d.; C *and* D *column*: 3 μm C$_{18}$-bonded silica gel, 10 cm × 4.6 mm i.d.; *eluent, 70% aqueous acetonitrile; flow rate* A *and* C, 1, B, 4, *and* D, 2 ml min^{-1}; *pressure drop* A, 1.2; B, 5.9; C, 13.4; D, 27 MPa; *detection: UV 254 nm. Peaks: 1, benzene; 2, acetophenone; 3, toluene, 4; naphthalene.*

column packed with 3 μm stationary phase material. However, a high flow rate through the 3 μm stationary phase material is limited by a high column back pressure. The separation could also be completed within 1.2 min on the short column packed with 3 μm stationary phase material by using a stronger eluent, as shown in Figure 1.2. Furthermore, the sensitivity was also improved by using the smaller-size stationary phase material because the sample is less spread out in the eluent and is more concentrated when it reaches the detector. The actual peak height in Figure 1.1C is 1.6 times that in Figure 1.1A. A small column packed with small particle-size stationary phase materials promises high performance and a high-speed separation both in theory and in practice. The following equation describes the relationship of the column length (L) to the column efficiency: $N = L/H$. The high plate number N required for good separation is proportional to the longer column length L and small H value. The term H is the height equivalent to a theoretical plate (HETP), which is the length of column needed to generate one theoretical plate. A good column has a high plate number for its length, and, thus, a good column has a low H value. The value of H can also be described by the following equation (which is described in detail in Chapter 5):

$$H = 2\lambda d_\mathrm{p} + \frac{2\gamma D_\mathrm{M}}{u} + \frac{q D_\mathrm{M} d_\mathrm{S}^2 u}{(1 + D_\mathrm{M})2D_\mathrm{S}} + \frac{d_\mathrm{p}^2 u}{D_\mathrm{M}}$$

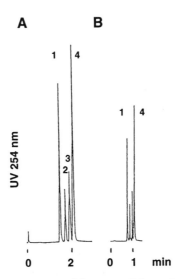

Figure 1.2 *Effect on eluent component of flow rate. Column: 3 μm C₁₈-bonded silica gel, 10 cm × 4.6 mm i.d.; eluent, 80% aqueous acetonitrile; flow rate A, 1, B, 2 ml min⁻¹; detection, UV 254 nm. Peaks: 1, benzene; 2, acetophenone; 3, toluene; 4, naphthalene.*
(Reproduced by permission from ref. 3)

This equation indicates that the particle size, d_p, is the main contributor to the H value. The smaller the particles, the higher the theoretical plate number. The optimum condition is obtained by the relationship between the theoretical plate height and the flow velocity.

2   Physical Considerations

High-speed separations can be achieved with a short column packed with 3 μm stationary phase material, as shown in Figure 1.2. The sensitivity was also improved by the use of smaller-size stationary phase materials, due to less sample diffusion inside the column. The following conditions are required to obtain such a separation.

    a. small-diameter, spherical stationary phase materials that have high physical strength;
    b. a high pressure pump with controlled flow rate;
    c. a system that limits sample diffusion, by considering the column design, using small inner diameter connecting tubing, and a small volume detector flow cell; and
    d. a detector and recorder capable of a high-speed response.

The theoretical plate number N of peak B can be calculated from the chromatogram given in Figure 1.3 by the following equation:

$$N = 16\left(\frac{V_R}{w}\right)^2$$

where V_R is the retention volume and w is the peak width at the base (measured in volume units). However, the retention volume includes the hold-up volume V_M (also called dead volume). The hold-up volume is the sum of the void volume of the column ($V_0 = YA$), the volume of the injector (OX) and the volume of the detector and connecting tubing (XY) as shown in Figure 1.3. The actual separation efficiency is defined as the effective theoretical plate number N_{eff}, which excludes the hold-up volume:

$$N_{eff} = 16\left(\frac{V_R - V_M}{w}\right)^2$$

Commercial instruments have a reasonable balance between the recommended column size and the volume of the column and connecting tubing (XY). However, the theoretical plate number of a single column may give different values on different instruments, and even on replacement of the components and parts of a single instrument. Such discrepancies can be understood in terms of differences in the mechanics of the instruments and the design of their parts.

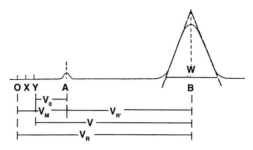

Figure 1.3 *A schematic chromatogram: V_R, retention volume; $V_{R'}$, adjusted retention volume; V, elution volume of peak; V_0, void volume; W, peak width; V_M, hold-up volume; OX, volume of injector; XY, volume of detector, including volume of tubing.*

The normally acceptable extra-column dead volume (OY in Figure 1.3) before there is a significant effect on the efficiency of a 15 cm long, 4.6 mm i.d. column should be less than 100 μl. This volume has to be reduced to less than 30 μl for a 5 cm long, 4.6 mm i.d. column. Replacement of the connecting tubing with shorter lengths of narrow-bore tubing and the selection of a smaller volume detector flow cell are necessary when using a shorter or narrower column. These changes together with a smaller column enable reductions in the volume of eluent required and in the separation time. This approach is economical and environmentally friendly. However, the reduced hold-up volume becomes technically very critical in the handling of smaller columns. Details of the basic mechanisms and the design of instrumentation are described in Chapter 2, which also covers the similarities and differences of various instruments.

3 Chemical Influences on the Separation Factor

One column can be used for different types of liquid chromatography by changing the eluent components. As an example, a column packed with octadecyl-bonded silica gel has been used for size-exclusion liquid chromato-graphy with tetrahydrofuran (THF), normal-phase liquid chromatography with *n*-hexane, and reversed-phase liquid chromatography with aqueous acetonitrile. Examples of the chromatograms are shown in Figure 1.4.

The elution volumes of polystyrene and benzene in the size-exclusion mode were 0.98 and 1.78 ml, respectively (Figure 1.4A). This means that separations by molecular size can be achieved between 0.98 and 1.78 ml in this system. In the normal phase mode the elution volumes of octylbenzene and benzene were 1.98 and 2.08 ml, respectively, in *n*-hexane solution (Figure 1.4B). This type of chromatography is called adsorption or non-aqueous reversed-phase liquid chromatography. These are adsorption liquid chromatography and non-aqueous reversed-phase liquid chromatography. The elution order of the alkylbenzenes in the reversed-phase mode using acetonitrile was reversed

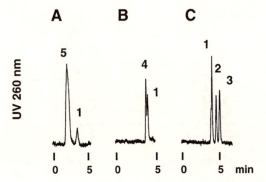

Figure 1.4 *Different modes of chromatographs using the same column. Column, 5 μm C$_{18}$-bonded silica gel, 15 cm × 4.6 mm i.d. Eluent A, tetrahydrofuran; B, n-hexane; C, acetonitrile; flow rate, 0.5 ml min^{-1} at ambient; detection, UV 260 nm. Peak 1, benzene; 2, ethylbenzene; 3, butylbenzene; 4, octylbenzene; and 5, polystyrene.*

(Figure 1.4C). The elution volumes of benzene and butylbenzene were 2.01 and 2.52 ml, respectively. The elution volumes became larger with the addition of water to the acetonitrile eluent. In each case the elution orders are based on the solubility of the solutes (except in size-exclusion liquid chromatography).

When separation cannot be achieved by improving the theoretical plate number of a column, it may be achieved by the selection of an appropriate stationary phase material and/or eluent. The degree of separation, the separation factor α, is the difference in retention volumes of analytes. The separation factor of two compounds is given by:

$$\alpha = \frac{V_{R2} - V_M}{V_{R1} - V_M}$$

where V_{R2} and V_{R1} are the retention volumes of peaks 2 and 1, respectively. These retention volumes depend on the properties of the solutes (analytes), stationary phase materials, and eluent components. A higher α value, *i.e.* an increase in the difference in retention volumes, can be achieved by using a different stationary phase material and/or eluent. Details of the selection of stationary phase materials and eluent are described in Chapters 3 and 4.

4   Basic Considerations of Liquid Chromatography

Identifying the most suitable separation conditions is the main objective of separation scientists. It is easier for skilled chromatographers, but this is a complicated subject for beginners. One approach is to find a chromatogram that exhibits the separation of a similar mixture. However, similar mixtures may have been separated under very different conditions; either the separation columns and/or the components of the eluent may have been different. Furthermore, the elution order is sometimes reversed. When an appropriate chromatogram is found in the literature, the conditions may need to be modified to take into account the other compounds in the mixture, any necessity for sample pre-treatment and the purpose of the separation. The chromatographer should be familiar with the capabilities and requirements of the following methods.

a. Pre-treatment of samples, stationary phase materials, and elution solvents;
b. Separations based on molecular size;
    (i)   aqueous size-exclusion liquid chromatography;
    (ii) non-aqueous size-exclusion liquid chromatography;
c. Normal-phase liquid chromatography;
d. Reversed-phase liquid chromatography;
e. Ion-exchange liquid chromatography;
f. Ion-pair liquid chromatography;
g. Chiral separation and affinity liquid chromatography.

Choosing the sample pre-treatment method is difficult. The most important consideration is the final condition of the target compounds. What kind of solution is obtained? The type of solvent and the concentration of the target compounds are very important in the selection of the separation conditions.

The pre-treatment of stationary phase materials is also important for silica gel and ion-exchangers, even when a new column is being used. Pre-treatment of the elution solvent is also important. High-performance liquid chromatography grade solvents from different manufacturers contain different amounts of impurities. The purity of the water is especially critical. The specified solvent for a special preparation stage is often not compatible with the desired chromatography. A large amount of such a solvent should first be injected, followed by measurement of the background of the chromatogram.

How are analytes retained on, or in, a stationary phase? This depends on the physicochemical interaction between the analytes and the stationary phase material. When a strong solvent, in which the solute readily dissolves, is used for elution the solute is eluted very quickly from the column. The forces holding an analyte on the stationary phase are similar to those responsible for dissolution in the solvent. Eight solubility properties are recognized: van der Waals (London dispersion) forces, dipole–dipole, ion–dipole, Coulombic and repulsion forces, charge-transfer complexation, and hydrogen-bonding and coordination bonds. The molecular interactions that are probably involved in retention in liquid chromatography can be explained by these interaction properties and are summarized in Table 1.1. The retention of a particular molecule is not due to only one property, but rather to a combination of several properties. The probable interactions can be estimated from the chemical structure of the analytes and stationary phase materials.

The separation conditions employed for size-exclusion liquid chromatography are simple. A strong solvent for analytes and a suitable stationary phase

Table 1.1 *Classification in liquid chromatographic methods*

Name of chromatography	Van der Waals	Repulsion	London dispersion	Hydrophobicity	Dipole–dipole	Charge transfer	Hydrogen bonding	Coulomb (ion–ion, ion–dipole)	Ligand formation	Complex formation	Salting-out	Steric effect
Size-exclusion	♦	○	○									
Reversed-phase	○	○	○	♦	△		△					
Reversed-phase ion-pair	○	○	○	♦	△			♦				
Ion-pair partition	○	○	○					○	♦			
Ion-exchange	○	○	○	○	○		△	♦		○		
Normal-phase	○	○	○		♦		○			△		○
Charge-transfer	○	○	○		○	♦		○				
Salting-out	○	○	○	○	○			○			♦	
Ligand-exchange	○	○	○					○	♦			
Chelation	○	○	○			○		○		♦		
Affinity	○	○	○	○	○	○	○	○	○	○		♦
Chiral separation	○	○	○	○	○	○				○		♦

♦ The most important mechanism, ○ important property, △ interaction depending on stationary phase material.

material are necessary. If the impurities have high relative molecular masses (M_r), size-exclusion chromatography can be used effectively. Size-exclusion liquid chromatography in conjunction with a recycling system can also separate isomers; however, it is time-consuming and the columns are usually expensive. If a mixture of molecules with a M_r of less than 2000 has to be separated and a recycling method seems to be insufficient for the separation, the following chromatographic technique can be carried out. If the sample concentration is large enough for chromatographic analysis, the eluted solution obtained by a size-exclusion chromatographic pre-treatment can be directly injected onto a liquid chromatograph using a syringe, after membrane filtration. If a good combination of stationary phase material and solvent cannot be found, then methods c–g in the above list are applied.

In reversed-phase liquid chromatography, increasing the molecular size increases the hydrophobicity of solutes and results in a greater retention volume. This indicates that the van der Waals volume is an important property in optimization. Increasing the number of substituents with π-electrons and hydrogen bonding increases the solubility in water, that is they increase the polarity of the solutes. This indicates that dipole–dipole and hydrogen-bonding interactions contribute to hydrophobicity. Therefore, these properties are important in controlling the retention volume in reversed-phase liquid chromatography. However, the π-electrons of stationary phase materials such as polystyrene gel and the hydrogen-bonding of non-endcapped bonded silica gels also contribute to the retention.

Many compounds can be analysed by methods c or d, and sometimes both methods c and d are employed. For a preparative-scale separation, method c (normal-phase chromatography) is suitable due to the easy removal of solvent. Identifying the best separation conditions for ion-exchange liquid chromatography (method e) is a little more difficult. Therefore, if the compounds can be separated using methods c or d, these methods should be used. Even saccharides, organic acids, and nucleic acids are often separated by methods c and d. The separation speed in ion-exchange liquid chromatography is also slower than that in normal and reversed-phase liquid chromatography.

Affinity liquid chromatography and chiral separations (enantiomer separations) require similar analyte properties. The solutes may have interactions through hydrogen-bonding, ligand formation, or Coulombic forces with the surface of stationary phase materials or the sites of additives; however, the selectivity is controlled by the steric effects of the structures of the analyte molecules and the recognition molecules (chiral selectors).

The physical and chemical properties of stationary phase materials are described in Chapter 3 (including methods for their synthesis) to clarify the differences in similar stationary phase materials supplied from different manufacturers. A detailed selection guide to solvents is given in Chapter 4. The unlimited selection of eluent components and their concentrations is a powerful force in developing separations in liquid chromatography. Although this area seems rather complicated, it is easy to understand the selection of a suitable eluent when you first identify the molecular properties of the analytes and solvents.

5 Bibliography

J.C. Giddings, 'Dynamics of Chromatography 1. Principles and Theory', Marcel Dekker, New York, 1965.

M.T. Gilbert, 'High-performance Liquid Chromatography', Wright, Bristol, 1987.

S. Hara, S. Mori, and T. Hanai, 'Chromatography. The Separation System', Maruzen, Tokyo, 1981.

H. Hatano and T. Hanai, 'New Experimental High Performance Liquid Chromatography', Kagaku Dojin, Kyoto, 1988.

E. Heftmann, 'Chromatography', 5th edn, Journal of Chromatography Library, Vol. 51A, Part A 'Fundamentals and Techniques', Elsevier, Amsterdam, 1992.

A.M. Krstulovic and P.R. Brown, 'Reversed-Phase High-Performance Liquid Chromatography', Wiley, New York, 1982.

S. Lindsay and D. Kealey, 'High Performance Liquid Chromatography', Wiley, Chichester, 1987.

V.R. Meyer, 'Practical High-Performance Liquid Chromatography', Wiley, Chichester, 1994.

P.A. Sewell, B. Clarke, and D. Kealey, 'Chromatographic Separations', Wiley, Chichester, 1987.

R.M. Smith, 'Retention and Selectivity in Liquid Chromatography', Journal of Chromatography Library, Elsevier, Amsterdam, 1995, Vol. 57.

L.R. Snyder and J.J. Kirkland, 'Introduction to Modern Liquid Chromatography', Wiley, New York, 1979.

L.R. Snyder, J.J. Kirkland, and J.L. Glajch, 'Practical HPLC Method Development', 2nd edn, Wiley, New York, 1997.

R.W. Yost, L.S. Ettre, and R.D. Conlon, 'Practical Liquid Chromatography, An Introduction', Perkin-Elmer, Norwalk, 1980.

CHAPTER 2

Instrumentation

The basic concepts of the instrumentation for liquid chromatography are described here, with the mechanism of their operation and their influence on the separation of analytes.

1 Chromatographic Systems

The components of all chromatographic systems are basically the same; however, the specifications and sizes differ between systems. A schematic diagram of a full-scale computerized system is shown in Figure 2.1. This type of system was originally proposed at the start of studies to design instrumentation for liquid chromatography, and the fully automated system is now commercially available. It is powerful, providing it is trouble-free. Its operation seems simple to the user, but trouble-shooting for this system is complicated. The basic concept of each of the components of the system is explained in the following sections.

2 Injectors

The maximum injection volume depends on the volume of the sample loop in the injection valve. The reproducibility of manual injection depends on the skill of the operator. The use of a small sample loop and an overflow injection of the sample solution so that the loop is fully flushed with sample are basic requirements for quantitative analysis. The highest injection reproducibility can be obtained by an auto-injector with a fixed sample loop. The smallest reasonable injection volume is 1 μl. A nl-scale injection valve can be constructed; however, the memory effect at the surface of contact parts affects quantitative analysis compared with the use of a μl-scale injection valve. For a semi-micro system, a low hold-up volume injection valve is desired. The minimum injection volume is 80 nl. For a preparative-scale injection, the sample loop can be easily replaced with a larger-volume loop, such as a 200 μl, instead of the standard 20 μl loop.

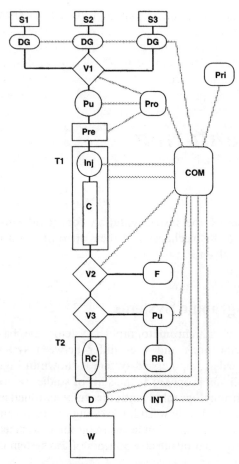

Figure 2.1 *Automated chromatograph. Components: solid line, liquid flow line; dotted line, communication line; S1, S2, and S3, solvent reservoirs; DG, degassing system; V1, V2, and V3, valves; Pu, pump; Pre, pressure sensor; Inj, injector; C, column; F, fraction collector; RC, reaction bath; RR, reaction reagent; D, detector; W, waste; T1 and T2, oven; COM, computer; Pro, solvent programmer; Pri, printer; INT, integrator or recorder.*

3 Eluent Delivery Systems

The pump is the heart of HPLC and must satisfy a number of requirements. It must deliver a constant and pulse-free flow (stability: less than $\pm 0.5\%$, or usually less than $\pm 1\ \mu l\ min^{-1}$) which must be reproducible under high pressures up to 50 MPa. The internal volume of the pump must be very small to enable rapid eluent changes and enable a precise gradient to be formed. The material of the flow lines must be compatible with all types of solvent, including organic solvents and strongly acidic and basic solutions. Four types of eluent delivery systems are popular. Widely used are the syringe pump and reciprocating piston pumps. A syringe pump is used for micro-scale liquid chromatography and capillary liquid/supercritical fluid chromatography. The most widely used is the double-piston reciprocating pump. This pump is suitable for a wide range of flow rates, from semi-micro liquid chromatography (LC) to preparative-scale LC under high pressure. The reciprocating piston pump design has been further modified in the master/slave reciprocating pump; one piston works to pump the eluent and the second piston works as the flow controller. The degree of pulsation of the flow depends on the mechanical and electrical capability of the manufacturer. The latest models provide improved precision of the flow rate, and can handle semi-micro-scale liquid chromatography. A precise flow rate ($\pm 1\ \mu l\ min^{-1}$) improves the reproducibility of gradient elution on standard size columns (6 mm > column i.d. > 3 mm). With the alternative diaphragm pump, the maximum pressure is less than that of the reciprocating pump, but the diaphragm resists a wider variety of solvents. A pneumatic amplifier pump is suitable for high flow rate operations such as preparative-scale liquid chromatography, but may not deliver a constant flow rate as it operates at constant pressure. This type of pump is also used for column packing.

Several methods are applied to reduce the separation time. The best way is the selection of a suitable column and an eluent using isocratic elution. However, much skill and knowledge are required to make such a system. A flow rate gradient, step-wise elution, or eluent composition gradients are commonly applied to reduce separation times.

Composition Gradient Delivery Systems

High and low pressure gradient systems are available, which mix two eluents to give a defined increase in eluent strength through a separation run, effectively reducing the separation time. A low pressure gradient system requires the pre-mixing of two or more solvents before pumping. This system is economical, but the internal system volume is higher than that of the high pressure gradient system. The precision of the high pressure gradient system, in which two pumps deliver the eluent components at changing rates to provide a total constant flow with a changing composition, is better but the cost is higher. The change and the reduction in elution time from an isocratic to a gradient elution are shown in Figures 2.2 and 2.3. The maintenance of a gradient elution system is more complicated than one using isocratic elution, but gradient elution is necessary for the separation of complex mixtures, such as amino acids, nucleic acids, and many biological samples.

Pressure Drop Changes due to Changes in the Eluent Composition

During a gradient elution the operational back-pressure changes due to changes in the viscosity as the eluent mixture changes[1] (Figure 2.4). When methanol and water are used as the components, the back pressure drop can become particularly high and sometimes stops the pump because it exceeds its maximum pressure limit. A high back pressure must be prevented, particularly when a longer column packed with small particle size is being used.

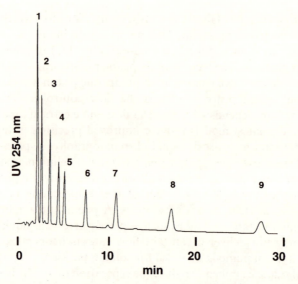

Figure 2.2 *Separation of aromatic compounds using isocratic elution. Conditions: column, 5 μm C_{18}-bonded silica gel, 15 cm × 4.6 mm i.d.; eluent, 0.001 M phosphoric acid in 55% aqueous acetonitrile; flow rate, 1 ml min^{-1}; temperature, ambient; detection, UV 254 nm. Peaks: 1, phenol; 2, 4-methylphenol; 3, 2,4-dimethylphenol; 4, 2,3,5-trimethylphenol; 5, benzene; 6, toluene; 7, ethylbenzene; 8, propylbenzene; and 9, butylbenzene.*

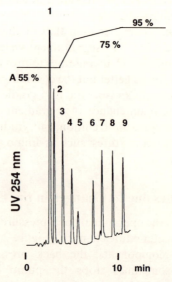

Figure 2.3 *Gradient elution of aromatic compounds. Chromatographic conditions as Figure 2.2 except for acetonitrile concentration.*

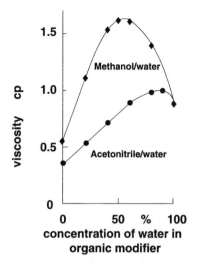

Figure 2.4   *Relationship of viscosity to water and organic modifier mixture ratio*

Flow Rate Gradients

Increasing the flow rate also reduces the separation time. However, this method is limited by the physical strength of the instrument and by the loss of detection sensitivity. The chromatograms of aromatic compounds separated isocratically and with a flow rate gradient are shown in Figures 2.2 and 2.5, respectively. Doubling the flow rate roughly halved the separation time; however, the peak areas became smaller. Flow programming elution is suitable for the separation of a mixture of homologous compounds.

Step-wise Elution

If the separation by isocratic elution is perfect, the eluent can be switched during the run to a stronger one to speed up late components (Figure 2.6). The peak height of longer-retained compounds increases and the retention time is shorter compared with the chromatogram of isocratic elution (Figure 2.2). This method is not suitable for quick repeat injections, due to the requirement for a significant column conditioning time after returning to the original conditions.

Recycle Elution

The recycle elution method can be applied to mixtures of very similar compounds that cannot be fully separated by a single pass through the column. This method makes more effective use of a column. The effluent from the column is repeatedly re-passed through the same column. The number of cycles multiplies the total theoretical plate number of the column if the system is

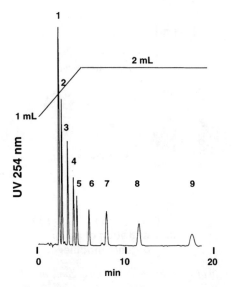

Figure 2.5 *Flow rate programming elution of aromatic compounds. Experimental con-*
ditions as Figure 2.2 except the flow rate. Flow rate increased from 1 to
2 ml min^{-1}.

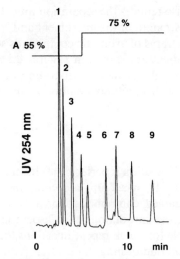

Figure 2.6 *Step-wise elution of aromatic compounds. Experimental conditions as in*
Figure 2.2 except the acetonitrile concentration was changed from 55 to 75%
in water.

carefully assembled. This method is usually applied to larger columns and
preparative-scale separations. Figure 2.7 shows the recycle chromatogram of
the purification of a mixture of anthracycline antibiotics, in which a fraction (or
heart-cut) of each separation is recycled through the column, becoming
successively more resolved on each cycle.

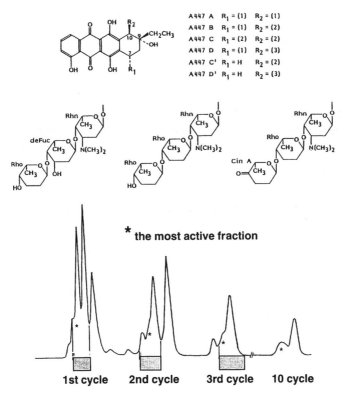

Figure 2.7 *Purification of a mixture of anthracycline antibiotics using heart cut recycle liquid chromatography to yield the most active fraction. Conditions: column, JAIGEL-310; eluent, chloroform–methanol–25% NH₄OH (200:5:1); flow rate, 4 ml min⁻¹; detection: refractivity index.*
(Reproduced by permission from Japan Analytical Industry data)

Column Switching Separation

The selectivity of different stationary phase materials can be applied using columns in sequence to provide high-speed isocratic separations instead of gradient elution. An example for amino acids analysis is shown later in Figure 4.15, where the same eluent was used for all of the separations and the fraction containing the sample components of interest was switched from one column to another.

4 Degassing Methods

Degassing of the eluent is important for trouble-free operation and highly sensitive detection,[2] otherwise the eluent may become supersaturated with air that is released as bubbles in the pump check valves or the detector flow cell. Several methods have been proposed, and two on-line systems have been commercialized. One method is bubbling helium gas into the eluent. Another is an on-line vacuum. Both mechanisms are simple, but the latter system is popular due to its compact size, low hold-up volume, and because it does not require expensive helium gas. Even if the eluent is a premixed mixture of water and organic solvent, pre-degassing is still important. The eluent container should be degassed using a water pump vacuum. If the operation is performed in an ultrasonic bath, degassing is completed within 1 min. Longer degassing can cause loss of the more volatile eluent components.

5   Column Temperature Control

If good reproducibility of the retention times is required, the column temperature should be kept constant. A water bath or an air oven are common systems, and thermo-tape is also effective if the column is made of steel. When temperature control is extremely important, pre-heating of the eluent is necessary. If the whole apparatus is placed in a temperature-controlled box, the temperature is constant, but such a system is not usually required.

6   Detectors

One of the first steps in the development of modern liquid chromatography was the automation brought about by adding an on-line detector. Subsequently, the development of small particle size high-performance stationary phase materials provided more efficient separation systems, which then became known as high-pressure or high-performance liquid chromatography (HPLC). A variety of analytical instruments have been employed as detectors. Originally, standard instruments were modified to monitor effluents from a high-performance column. The highest possible sensitivity and selectivity of detectors are required in biomedical research, and led to the introduction of dedicated instruments with high qualitative performance. The basic mechanism is the same as that of the original instruments, but technical improvements have been made in the mechanics and electronics. The specifications of modern detectors for HPLC are summarized in Table 2.1.

Spectroscopic Detectors

Energy from a light beam is absorbed by molecules with a chromophore. An absorption spectrophotometer uses this mechanism, and the energy loss depends on the concentration and molecular absorption constant of analyte molecules and the wavelength of the light. The most popular detector, the

Table 2.1 *Specificity of detectors*

Detector	Sensitivity	Cell volume μl	Remarks
Ultraviolet (visible) spectroscopy	1 ng ml^{-1} 1 × 10^{-3} AUFS	4.5–20	Common detector; selection of wavelength improves the accuracy of quantitative analysis, wavelength scanning or photodiode array permits qualitative analysis. W-lamp or complex lamp is suitable for visible region detection
Fluorescence spectroscopy	10 pg ml^{-1} 1 × 10^{-4} FS	7–60	Sensitive for fluorescent compounds
Refractive Index	10 ng ml^{-1} 1.6 × 10^{-7} RIU	5–14	All purpose; temperature control for high sensitive detection, usually used for preparative LC or size exclusion LC, eluent must be isocratic
Polarimeter	20 μg ml^{-1}		Specific for optical isomers
Infrared spectroscopy	10 ng ml^{-1}		Suitable for carbonyl compounds; NaCl, CaF$_2$ and sapphire cells are available
Conductivity	10 ng ml^{-1}	1.5	Detection limit 0.001 μmho is equivalent to 0.01 ppm NaCl, suitable for ion detection
Polarography	1 ng ml^{-1}, 2 × 10^{-8} μA	10	Suitable for ions, fatty acids, saccharides
Coulometry	50 pg s^{-1}, 5 × 10^{-8} A		
Amperometry	50 ng ml^{-1}, 10 pA		Suitable for catecholamines, phenols
Pulsed amperometry	50 ng ml^{-1}, 10 pA		Suitable for saccharides
Flame ionization	1 ng ml^{-1}, 10 pA		All purpose; suitable for saccharides and non-volatile compounds
Electron capture	100 pg ml^{-1}		Suitable for halogenated, phosphorus and oxygenated compounds
Mass spectrometer	10 pg ml^{-1}		All purpose; qualitative and quantitative detector
Inductive coupled plasma	20 ng ml^{-1}		Element selective
Atomic absorption	20 pg ml^{-1}		Organometallic compounds
Reaction heat	ng s^{-1}	9	All purpose
Radioactivity	^{14}C:2nCi, ^{3}H:5nCi	50–700	Metabolite analysis

Refs: MS: *Anal. Chem.*, 1981, **53**, 1603; ICP: *J. Chromatogr. Sci.*, 1985, **23**, 4; AAS, *Analyst*, 1981, **106**, 921.

ultraviolet absorption detector, uses a deuterium lamp as the light source. A visible region absorption detector requires a tungsten lamp. A combined lamp is useful for monitoring both in the ultraviolet and visible regions. Pre- or post-column derivatization of sample molecules to introduce a strong chromophore can be applied for highly sensitive and selective detection. Pre-column derivatization is also used to aid selective extraction from the effluent for collection purposes.

Fluorescence detection is usually more sensitive than absorption detection, but the number of naturally fluorescent compounds is limited. Pre- or post-column derivatization can also be applied for this type of detection. Chemiluminescence detection is the most sensitive method for some fluorescent compounds.

A light beam is refracted to different extents by different compounds. This mechanism is used for refractive index detection. This detector is not sensitive and the selectivity differences are negligible for homologous compounds, but any solvent with a different refractive index to the analyte can be used as the eluent. This detector is mainly applied to size-exclusion and preparative-scale liquid chromatography.

Infrared spectroscopy is often used for qualitative analysis, and its powerful selectivity means that it can be used as a detector. However, the absorption of the eluent molecules, particularly in reversed-phase separations, often interferes with the detection of analytes. The infrared absorption detector therefore requires mechanical assistance to eliminate the solvent or needs powerful computer assistance to eliminate the background signal.

Atomic absorption and inductively coupled plasma spectrometers are metal-selective spectrometers used for organic metal analysis. The connection of these spectroscopic instruments to a liquid chromatograph is relatively simple. Chromatograms of alkylmercury[3] and aminoplatinum analytes[4] are shown in Figures 2.8 and 2.9, respectively.

A polarimeter can be used for the identification of optically active compounds,[5] as shown in Figure 2.10.

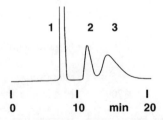

Figure 2.8 *Detection of alkylmercury compounds using flameless atomic absorption detector. Conditions: column, Corasil I, 50 cm × 2.1 mm i.d.; eluent, n-hexane; flow rate, 0.5 ml min⁻¹; detection, flameless atomic absorption spectrometer. Peaks: 1, benzene; 2, ethylmercury chloride; and 3, methylmercury chloride.*

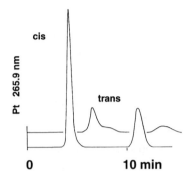

Figure 2.9 *Inductively coupled plasma detection of diaminodichloroplatinum(II). Conditions: column, Shodex OH, 25 cm × 4.1 mm i.d.; eluent, 0.01 M phosphoric acid; flow rate, 1 ml min⁻¹; detector, emission at platinum line 265.9 nm. Samples: cis and trans diaminodichloroplatinums. The second peaks are considered to be their oligomers.*

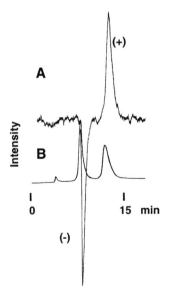

Figure 2.10 *Chromatograms of (±)-trans-stilbene oxide. Conditions: column, Chiralpak OP(+), 25 cm × 4.6 mm i.d.; eluent, methanol; column temperature, ambient; flow rate, 1 ml min⁻¹; detector A, polarimeter; B, refractive index.*

Electrochemical Detectors

Electrochemical detectors, which are based on the electrochemical oxidation or reduction of the analyte, can be applied to the analysis of selected compounds such as phenols. It is physically simple, but is very sensitive for catecholamines. However, the adsorption of reacted molecules on the surface of the electrodes can reduce the conductivity. To overcome this problem a pulsed voltage is applied, which cleans the electrode surface between measurements. This pulsed amperometric detection is also sensitive for carbohydrates.

Conductivity detection, which involves measuring changes in the conductivity of an aqueous solution between two electrodes, is employed in ion chromatography for the detection of ionized analytes.

Mass Spectrometer

The mass spectrometer is a very sensitive and selective instrument. However, the introduction of the eluent into the vacuum chamber and the resulting significant pressure drop reduces the sensitivity. The gas exhaust power of a normal vacuum pump is some 10 ml min^{-1} so high capacity or turbo vacuum pumps are usually needed. The gas-phase volume corresponding to 1 ml of liquid is 176 ml for n-hexane, 384 ml for ethanol, 429 ml for acetonitrile, 554 ml for methanol, and 1245 ml for water under standard conditions (0 °C, 1 atmosphere). The elimination of the mobile phase solvent is therefore important, otherwise the expanding eluent will destroy the vacuum in the detector. Several methods to accomplish this have been developed. The commercialized interfaces are thermo-spray, moving-belt, electrospray ionization, ion-spray, and atmospheric pressure ionization. The influence of the eluent is very complex, and the modification of eluent components and the selection of an interface are therefore important. Micro-liquid chromatography is suitable for this detector, due to its very small flow rate (usually only $10 \text{ } \mu\text{l min}^{-1}$).

Other Detectors

Flame ionization and electron capture detectors, which are used in gas chromatography, have been modified for liquid chromatography but have not been widely used. It is again necessary to eliminate the eluent to improve the sensitivity and to broaden their application. Moving-wire and moving-belt flame ionization detectors are useful for the detection of organic compounds, but a difference of greater than 80 °C between the boiling point of the analyte and the eluent is required to prevent loss of the analyte. The sensitivity of the electron capture detector is 0.12 ng ml^{-1} for aldrin; however, only volatile organic solvents can be used as the eluent components. A radioisotope detector has been used for radiolabelled metabolite analysis. Electron spin resonance (ESR) has been used for the analysis of radicals. Nuclear magnetic resonance spectroscopy (NMR) with a flow cell has also been used for qualitative analysis.

The development of high-power NMR (800 MHz) spectrometers permits the simple operation of LC-NMR for metabolite analysis.

Measurement of Detector Sensitivity

For ultraviolet and visible spectroscopic detectors, a standard solution of a compound whose molar absorption constant is known must be prepared, and placed in the flow cell. The absorbance obtained is then compared with the value measured by a standard spectrophotometer.

For a fluorescence detector, quinine sulfate is used as the standard compound. The flow cell is filled with a standard solution and the fluorescence intensity is measured. The value is compared with that measured by a fluorescence spectrophotometer. This standard solution is also used for fixing the wavelength and position of the flow cell. The Raman spectrum of water can also be used for this purpose.

For refractive index detectors, 13.33 g of sucrose is dissolved in 100 ml of pure water and then diluted 200 times. The refractive index of the final solution is 1×10^{-4} RI, which is used to calibrate the instrument.

7 Recording Systems

The visualization of the detector signal helps to clarify the separation conditions. Pen recorders were originally used, but the integrator is now popular due to the automatic reporting of both the retention time and peak area or height. The effective use of an integrator makes quantitative analysis easier. Computer-based integrators are powerful for the storage and further arrangement of data, and can also be used for column evaluation and as a system controller.

Selection of Detector Time Constant

The selection of the time constant of the detector and recording system is very important in high-speed separations. If the peak height monitored at a high flow rate is lower than that measured at the standard flow rate, what is the cause? Sometimes the peak height is lower than that expected theoretically. Carrying out a separation at a flow rate of 1 ml min^{-1} and a recorder time constant of 1 s gave chromatogram A in Figure 2.11, and at 4 ml min^{-1} chromatogram B was obtained. The peak heights in chromatogram B are *ca.* 60% of those of A. The peak heights in chromatogram C, using a time constant of 2 s and flow rate of 4 ml min^{-1}, are *ca.* 50% of those of A. When the time constant was then reduced to 0.1 s, chromatogram D was obtained with peak heights *ca.* 80% of those of A. This experiment demonstrates the importance of time constant selection. The importance can be further observed by examining the peak height changes of the first peak in Figure 2.11. These chromatograms were monitored on a chart recorder. When the experiments were monitored by an integrator, the area of peak 1 in Figure 2.11 was constant in all four runs, but the peak heights

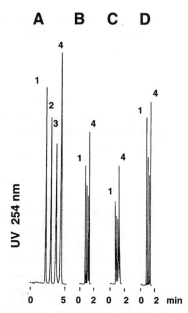

Figure 2.11   *Effect of time constant of detector. Conditions: column:* 10 μm C_{18} *silica gel,*
15 cm × 4.6 mm i.d.; *eluent, 70% aqueous acetonitrile; flow rate for* A,
1 ml min^{-1}, *for* B–D, 4 ml min^{-1}; *time constant* A *and* B, 1.0 s, C, 2.0 s, D,
0.1 s. *Peaks:* 1, *benzene*; 2, *acetophenone*; 3, *toluene*; *and* 4, *naphthalene*.

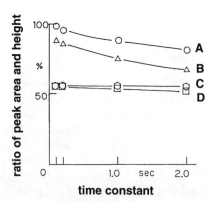

Figure 2.12   *Ratio of peak area and height related to time constant of detector. Experi-*
mental conditions are the same as those in Figure 2.11. Peak height and area
ratios are calculated from the data of peak nos. 1 and 4 as a % of the
maximum value. Lines: A, *peak height ratio of chromatograms measured at*
1.0 ml min^{-1} *flow rate*; B, *peak height ratio of chromatograms measured at*
4.0 ml min^{-1} *flow rate*; C, *peak area ratio of chromatograms measured*
at 1.0 ml min^{-1} *flow rate*; *and* D, *peak area ratio of chromatograms*
measured at 4.0 ml min^{-1} *flow rate*.

were affected by the time constant. The peak area ratio between peaks 1 and 4 was almost constant from 1 to 4 ml min^{-1} (Figure 2.12, lines C and D), while the time constant was varied from 0.1 to 2 s. A rapid time constant is required when the flow rate exceeds 4 ml min^{-1}. The peak height ratio of peaks 1 and 4 was affected by the time constant (Figure 2.12, lines A and B); therefore, the selection of the time constant for quantitative analysis is important. Furthermore, the optimum peak recognition parameters of an integrator should be determined for fast separations. When the time constant of the detector and the peak recognition parameter are too small, the noise level becomes high and quantitative analysis becomes difficult. Suitable conditions must be selected in each case for quantitative analysis.

8 Columns and Connectors

The best material for column tubing is glass, but careful handling is required due to its physical vulnerability. The most commonly used column material is stainless steel tubing, and the inner surface is polished like a mirror. These columns are physically stable and are compatible with a variety of compounds and eluent components. Some biological materials interact with metals; therefore, organic polymer and glass-lined stainless steel tubes are preferable for biological samples, and also for eluents with high ionic strengths which can attack the steel. The design of the inlet and outlet connections of columns depends on the physical strength of the materials. However, different connectors and fittings from different systems should not be mixed. Different connectors sometimes damage the efficiency of a whole system because they can create significant extra column dead volumes. The best approach is to use one connection system, even with instruments from different manufacturers. Replacement of the connection system is now simple, and does not damage individual components. Finger-tight polymer-type (PEEK) connectors are desirable if the instrument is operated at usual pressures (up to 20 MPa).

Column systems are basically classified into two groups: conventional and cartridge columns. A conventional column usually demonstrates better column efficiency than does a cartridge column due to a higher column packing pressure, but its usefulness depends on the quality control system of the manufacturer and the skill of their technicians. A cartridge column is economical, but its theoretical performance and lifetime can be less than a conventional column.

9 Flow Cell Volume and Connecting Tube Dimensions for High Efficiency Operation

If the correctly sized flow cell and connecting tubing are not used, the high efficiency of a column or high theoretical plate number columns cannot be effectively used. The detector cell volume contributes hold-up volume. The larger is the cell volume, the greater the peak broadening. The cell volume should be less than 10% of the typical volume required to elute a peak. The peak volume is related to the retention factor k, and the longer the retention time, the larger is the peak volume. The cell volume is thus particularly critical for compounds with short retention times. A well-designed small volume flow cell is also required for smaller diameter or short columns. The relationship between the maximum desirable cell volume and column i.d. of a 10 cm long column is given by:

$$\text{Cell volume } (\mu l) = 4.2 \times 10^{-4}x^2 - 3.6 \times 10^{-6}x + 8.4 \times 10^{-5}$$

for a compound with $k = 1$, where x is the column i.d.

The connecting tubing should be as short and narrow as possible. The volume of a 20 cm × 0.5 mm i.d. tube is 39.3 μl. That of a 20 cm × 0.25 mm i.d. tube is 9.8 μl, and that of a 20 cm × 0.125 mm i.d. tube is 2.5 μl. Some detectors are equipped with a heat exchanger that consists of a metal block containing a capillary tube. The volume of this tube also affects the theoretical plate number. If highly sensitive operation is not required, the heat exchanger can be removed or bypassed.

10   Other Components in a Liquid Chromatograph

A fraction collector and a post-column derivatization system were included (Figure 2.1) for a comprehensive and multi-purpose instrument. However, the fraction collector is needed only when collecting components from the effluent, and is generally not included in an analytical system. The post-column derivatization system is connected only when required for the selective and sensitive detection of specially targeted compounds. Usually, most compounds are directly detected by an on-line spectroscopic or other detector.

A computer (Figure 2.1) can handle the complicated operations of controlling the auto-injector (injecting different volumes, adding an internal standard compound and selecting sample vials), the pumps (controlling the flow rate, selecting the different gradient methods, and changing the eluent composition), the detectors (selecting wavelength, sensitivity, time constant, and the optimization), the column temperature, the recording system, and other parts of the system. Additional software can improve the quality of reports, including the modification of chromatograms and the results of quantitative analysis. Furthermore, combination with a computational chemical analysis system can enable the production of quantitative structure–retention relationships.

11   Trouble-shooting

General problems and the appropriate maintenance procedures are summarized in Table 2.2. The details of individual instruments are well described in the manufacturers' manuals. The common problems are due to poor maintenance of the instruments and poor understanding of the specificity of stationary phase

Table 2.2 *Trouble shooting and maintenance*

Trouble: Diagnosis	Maintenance
1. No response after switching on	
A. Fuse	Replace fuse
B. Electric cable	Check electric cable
2. No eluent flow from pump	
A. No movement of pump	
a. No current	See 1A and 1B
b. Lack of grease oil	Oil
c. Safety pressure regulator on	Fix or replace pressure gauge
d. Gears jammed	Change flow rate; if this does not work, then call the manufacturer
e. Recrystallized salt	Wash with water; call manufacturer
B. Pump piston moves	
a. Leak in line	Fix or replace appropriate part
b. Air bubble in pump	Degas the eluent, disconnect the pump and remove air by using a high flow rate
c. Dirty check valve	Disconnect pump, wash with a high flow rate or remove check valve and wash in an ultrasonic bath with a strong solvent; finally, overhaul the check valve
d. Aged piston seal	Fit new seal. Call manufacturer if using a syringe or diaphragm type pump
C. Back pressure too high	
a. Plug in column, precolumn, guard column, or on-line filter due to dust from injector rotor seal or sample	Clean inlet side of column or replace filter
b. Expansion of stationary phase	Wash column using intermediate solvent at low flow rate; change eluent; change chromatography system
c. Solvent filter blocked	Replace with a suitable one
3. Eluent flow present	
A. Low flow rate	
a. 2B a–d	
B. No response from detector	
a. No power to detector	1A, 1B
b. Aged lamp or photo cell	Replace lamp or photo cell
c. Others	Call manufacturer
C. No recorder movement	
a. No current	1A, 1B
b. Broken wire	Replace
c. Malfunction of resistance wire	Clean; if this fails, go to f.
d. Selection of polarity	
e. Flow cells unbalanced	Clean cell or replace solvent
f. Others	Use another recorder, and call manufacturer

(*continued*)

Table 2.2 *Continued*

Trouble: Diagnosis	Maintenance
4. Flow present and recorder responding	
A. Irregular noise on baseline	
a. Air bubble in flow cell	Degas eluent; remove air bubble from flow cell
b. Small bubble or immiscible solvent in flow cell	Wash flow cell with ethanol, THF or 6 M HNO_3
c. Dirty flow cell window	Apply b above; clean flow cell
d. Insufficient light energy	Replace lamp; if using an unsuitable eluent component of absorbance at detection wavelength, change eluent
e. Malfunction of ground/earth	Fix or change earth connection
B. Drifting baseline on output	
a. Unstable current or voltage	Use stabilizer on power line
b. Dust on light line	Clean
c. Leak at flow cell	Fix it, or replace seal
d. Using refractive index detector	Flow rate too high; drafts; change prism position
e. Step-wise movement	Clean recorder slide wire; slightly increase gain of recorder
f. Vibration of pen recorder	Decrease recorder gain a little
C. Baseline drift without gradient elution	
a. Dirty flow cell, expanding air bubble	Wash flow cell
b. Using refractive index detector	Control detector temperature
c. Aged reference battery of recorder	Replace battery in recorder
d. Soluble stationary phase	Change chromatographic system
e. Unstable current or voltage	B a
f. Expanding air bubble inside flow cell	Remove air from line, especially from pump head; degas eluent
D. Signal out of range of baseline	
a. 3B c	3B, C
b. 4A b,c	4A b,c
c. Malfunction of light beam in detector	Adjust flow cell position; remove flow cell and fix it
d. Solvent	Use HPLC grade solvent; use freshly prepared eluent
e. Reference flow cell	Fill with the same eluent
f. Fresnel type refractive index detector	Adjust zero balance
5. Bad peak reproducibility even when conditions 1–4 are satisfied	
A. Increased peak retention times	
a. 3A	3A
b. Column temperature reduced	Use fixed oven temperature
c. Unstable column	If due to previous solvent, wash column sufficiently, if due to stationary phase, wash column sufficiently; if stationary phase is soluble in the eluent, change chromatographic system
d. Chart speed	Fix recorder

(continued)

Table 2.2 *Continued*

Trouble: Diagnosis	Maintenance
B. Decreased peak retention times	
a. Aged stationary phase	Replace stationary phase or column; if stationary phase soluble, change chromatographic system
b. Remains of previous eluent (especially gradient elution)	Wash column sufficiently
c. Increasing column temperature	Fix oven temperature
C. Resolution becomes poor	
a. 5B a–c	5B a–c
b. Unbalanced density of column stationary phase	Repack column
c. Overloading of sample	Resolution is poor even if repeated under initial conditions, see 5C, a,b
D. Poor peak area reproducibility	
a. Poor injection	Wash micro injection syringe; fix injector leakage
b. Trouble with detector	3B, b; 4A, c; 4C, b; 4D, f; if non-linear range of detector used, change concentration
c. Trouble with recorder	3C, b,c; 4C, c; fix the gain of the recorder
6. Tailing peak	
A. Polluted column	5A
7. Leading peak	
A. Unsuitable chromatographic system	Change chromatographic system

materials and solvents. A high-performance liquid chromatograph may be used by many operators under different conditions. All operators should therefore study the maintenance manuals and keep a stock of frequently required replacement parts, such as rotor seals and pump piston seals. Many toxic organic solvents and eluent components are used in HPLC, and operators should therefore use safety glasses and work in areas that have adequate ventilation.

12 References

1. H. Colin, J. C. Diez-Mesa, G. Guiochon, T. Czajkowska, and I. Miedziak, *J. Chromatogr.*, 1978, **167**, 41.
2. A. Nagai, in 'Advances in Liquid Chromatography', eds. T. Hanai and H. Hatano, World Science, Singapore, 1996, pp. 151–167.
3. W. Funasaka, T. Hanai, and K. Fujimura, *J. Chromatogr. Sci.*, 1974, **12**, 517.
4. T. Hanai, J. Hubert, G. Luissier, M. Bellavance, and M. Lefebvre, 'Proceedings of 4th International Symposium on Column Liquid Chromatography', Boston, May 1979.
5. T. Hanai, I. Suzuki, and K. Nakanishi, *Anal. Sci.*, 1985, **1**, 483.

Reference Books

J.W. Dolan and L.R. Snyder (eds.), 'Troubleshooting LC Systems', Humana Press, Clifton, 1989.

E. Katz (ed.), 'Quantitative Analysis using Chromatographic Techniques', Wiley, Chichester, 1987.

E.S. Yeung (ed.), 'Detectors for Liquid Chromatography', Wiley, New York, 1986.

H. Parvez, M. Bastart-Malsot, S. Parvez, T. Nagatsu, and G. Carpentier (eds.), 'Electrochemical Detection in Medicine and Chemistry', VNU Science Press, Utrecht, 1987.

R.W. Frei and K. Zech (eds.), 'Selective Sample Handling and Detection in High-Performance Liquid Chromatography', Elsevier, Amsterdam, 1988.

K. Blau and J.M. Halket (eds.), 'Handbook of Derivatives for Chromatography', 2nd edn, Wiley, Chichester, 1993.

T. Hanai and H. Hatano (eds.), 'Advances in Liquid Chromatography', World Science, Singapore, 1996.

CHAPTER 3

Preparation, Testing, and Selectivity of Stationary Phase Materials

The selection of the stationary phase material is generally not difficult when the retention mechanism of the intended separation is understood. Stationary phase materials have different physical and chemical properties that can be understood from their chemical structure. They can be classified according to their retention mechanism and chemical structure (Table 3.1). The symbol ♦ indicates a strong interaction. All stationary phase materials can be classified in this manner, including chiral and affinity phases. For example, alkyl- or phenyl-bonded vinyl alcohol gels are used in reversed-phase liquid chromatography, and ion-exchange group-bonded vinyl alcohol gels can be used as cation or anion-exchangers. Non-bonded materials can be used for size-exclusion liquid chromatography.

The fundamental behaviour of stationary phase materials is related to their solubility–interaction properties. A hydrophobic phase acts as a partner to a hydrophobic interaction. An ionic phase acts as a partner for ion–ion interactions, and surface metal ions as a partner for ligand complex formation. A chiral phase partners chiral recognition, and specific three-dimensional phases partner affinity interactions.

Stationary phase materials are synthesized from different raw materials. Those stationary phase materials synthesized from inorganic materials, such as silica and alumina, are physically strong but chemically unstable. Conversely, stationary phase materials synthesized from organic materials, such as polystyrene or poly(vinyl alcohol), are chemically stable but physically weaker. Improvements in the chemical stability of inorganic stationary phase materials and in the physical strength of organic stationary phase materials are required; the marketed products do not have both and have to be used under restricted conditions in liquid chromatography.

Table 3.1 *Stationary phase material selection guide*

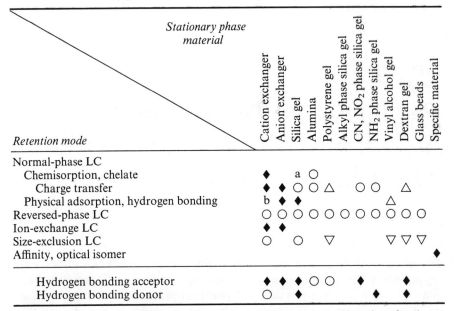

Retention mode	Cation exchanger	Anion exchanger	Silica gel	Alumina	Polystyrene gel	Alkyl phase silica gel	CN, NO$_2$ phase silica gel	NH$_2$ phase silica gel	Vinyl alcohol gel	Dextran gel	Glass beads	Specific material
Normal-phase LC												
Chemisorption, chelate	♦		a	○								
Charge transfer	♦	♦	○	○	△		○	○		△		
Physical adsorption, hydrogen bonding	b	♦	♦						△			
Reversed-phase LC	○	○	○	○	○	○	○	○	○	○	○	
Ion-exchange LC	♦	♦										
Size-exclusion LC	○		○		▽				▽	▽	▽	
Affinity, optical isomer												♦
Hydrogen bonding acceptor	♦	♦	♦				○	○	♦		♦	
Hydrogen bonding donor	○	♦	♦						♦		♦	

♦: strong interaction; ○: possible interaction, △: weak interaction; ▽: a wide variety of stationary phase materials available; a Ferrous ion in silica; b especially COOH form of ion-exchanger.

1 Synthesis of Stationary Phase Materials

The selectivity of stationary phase materials can be understood if the method of their synthesis is understood. Differences in the same type of stationary phase material from different manufacturers or even from the same manufacturer depend on the synthetic methods and the quality control that has been employed. Details of the individual synthetic processes from different manufacturers have not been published, but are basically the same.[1,2]

Silica Gel

Silica gel is a three-dimensional polymer of silicic acid, usually synthesized from tetrachlorosilane or sodium silicic acid. The reaction is as follows:

either $SiCl_4 + 2H_2O + 2Ag_2O \rightarrow 4AgCl + Si(OH)_4$

or $NaSiO_4 + 4(H^+\text{-form ion-exchanger}) \xrightarrow{\text{ion-exchange}}$
$$Si(OH)_4 + 4(Na^+\text{-form ion-exchanger})$$

The removal of sodium and other metal ions is very important to obtain pure silica gels. The next step in both routes is:

$$2Si(OH)_4 \rightarrow (OH)_3SiOSi(OH)_3 + H_2O \rightarrow \text{silica gel}$$

Silica gel is also made from colloidal silica (silica sol):

$$Si(OH)_2{}^{6-} \rightarrow Si(OH)_4 \rightarrow \text{silica gel}$$

The particle diameter and pore size of the silica gel materials depend on the concentration of silicic acid, the pH, the solvent and mixing conditions, and the reaction temperature. Technical skill is necessary for reproducible large-scale synthesis.

The purity of silica gel made from sodium silicate depends on the ion-exchange stage, and is usually about 99%. A purer silica gel can be produced from organosilica compounds. In this method tetraethoxysilane is polymerized to polyethoxysilane. The emulsion of polyethoxysilane in a water–alcohol mixture is further dehydroxylated and polymerized with a catalyst. The hydrolysed silica gel is sieved, dried, and sieved again before use as a stationary phase material. The pore size and particle size depend on the concentration of polyethoxysilane, the solvent, the temperature, the amount of catalyst, and the mixing conditions. Strong agitation leads to a smaller particle size, and a large amount of catalyst leads to a large pore size. The purity of this silica gel is more than 99.99%, depending on the purity of the water used for washing and sieving.

The surface of the virgin silica gel is covered with water. Heating to 110 °C forms a monolayer of silanol groups on the surface (8 μmol of silanol groups per m^2). Heating to over 600 °C produces a surface siloxane structure, and this is rehydrated to the silanol form under humid conditions, as shown in Figure 3.1.

Electron microscopy photographs of classic irregular (A) and spherical (B) silica gels and high purity, spherical silica gel (C), whose purity is over 99.99%, are shown in Figure 3.2. The surface of pure silica gel is very smooth and physically very stable due to the homogeneous structure. Modification of the

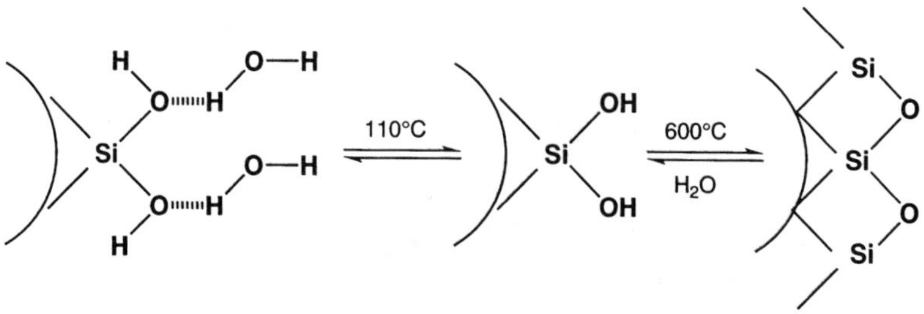

Figure 3.1   *Modification of silica gel surface.*

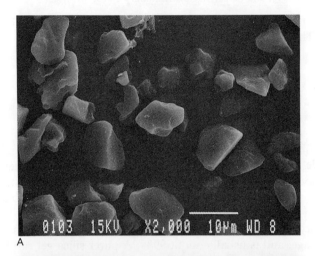

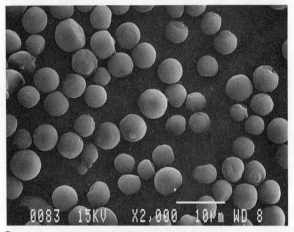

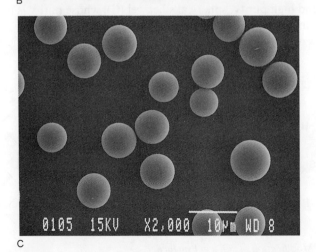

surface is easier (see section below on surface modification) than for ordinary silica gels.

Surface Modification of Silica Gel

Spherical porous silica gel is the easiest stationary phase material to handle; however, although it is physically strong it is chemically unstable. Surface modification can expand its capability for different modes of chromatography, such as normal-phase, reversed-phase, size-exclusion, and ion-exchange liquid chromatography. These stable modifications are performed by chemical derivatization of the surface silanol groups.

Modification Using a Halogenated Silica Surface. Silica gel is heated in 2 M HCl, filtered off, and washed with water; the dried silica gel is then reacted with pure thionyl chloride. After removal of the excess of thionyl chloride and by-products (SO_2 and HCl) under vacuum, chlorinated silica gel ($\equiv$SiCl) is obtained. This is unstable in water and alcohol. The chlorinated silica gel can then be alkylated by one of a number of different reactions:

$$\equiv Si\text{–}OH \rightarrow (SOCl_2, TiCl_4, \text{ or } SiCl_4) \rightarrow \equiv SiCl$$
$$\equiv SiCl \rightarrow (RMgCl) \rightarrow \equiv SiR + MgCl_2 \text{ (Grignard reaction)}$$
$$\equiv SiCl \rightarrow (RLi) \rightarrow \equiv SiR + LiCl$$
$$\equiv SiCl \rightarrow (RNH_2) \rightarrow \equiv SiNHR + HCl$$

The chlorinated silica gel suspended in diethyl ether, dimethyl sulfoxide, or dioxane will also react with diamine or amino compounds. After being washed in an excess of amine and HCl, and then dried under vacuum, the final stationary phase materials are obtained:

$$\equiv SiCl + H_2N\text{–}(CH_2)_n\text{–}X \rightarrow \equiv Si\text{–}NH\text{–}(CH_2)_n\text{–}X + HCl$$
$$X = CH_3, NH_2, CO_2H, SO_3H, CN, \text{ or } NO_2$$

Further modification of diamino bonded products with halogenated compounds can lead to a variety of stationary phase materials:

$$\equiv Si\text{–}NH\text{–}(CH_2)_n\text{–}NH_2 + Y\text{–}CH_2\text{–}R\text{–}X \rightarrow \equiv Si\text{–}NH\text{–}(CH_2)_n\text{–}NH\text{–}CH_2\text{–}R\text{–}X$$
$$R = CH_2, C_6H_4$$

Modification Using a Chlorosilane. Silica gel is heated in 2 M HCl, filtered off, and washed with water. The dried silica gel is then boiled under reflux in toluene

---

Figure 3.2   *Electron microscopic photos of silica gels:* A, *irregularly shaped;* B, *ordinary spherical; and* C, *high purity silica.*
(Contributed by Hideyuki Negishi, Kanagawa Dental College, Yokosuka, Japan)

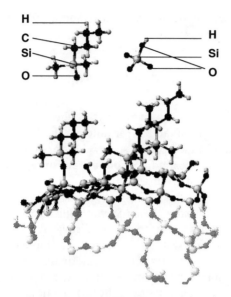

Figure 3.3 *Molecular model of butyl-bonded silica gel.*

containing the chlorosilyl reagent, using pyridine as the catalyst. Three types of chlorosilyl reagents are available:

$$3 \equiv\!Si\!-\!OH + RSiCl_3 \rightarrow \equiv\!SiO_3SiR + 3HCl$$
$$2 \equiv\!Si\!-\!OH + R_2SiCl_2 \rightarrow \equiv\!SiO_2SiR_2 + 2HCl$$
$$\equiv\!Si\!-\!OH + R_3SiCl \rightarrow \equiv\!SiOSiR_3 + HCl$$

A depiction of a butyl-bonded phase is shown in Figure 3.3. Monochlorosilane reagents produce only monomeric phases; however, trichlorosilane reagents can produce both monomeric and polymeric phases depending upon the concentration of silyl reagent and the surface area. It is difficult to make a multilayer bonded phase, even when a large quantity of trichlorosilane is used for the reaction.

Synthesis of Ion-exchanger Stationary Phases

Cation-exchanger. Phenyl-bonded silica gel is suspended in chlorosulfonic acid–acetone and boiled under reflux. After filtration and washing with acetone and dilute HCl, the sulfonated gel is then converted into a suitable salt form:

$$\equiv\!Si\!-\!Ph \rightarrow \equiv\!Si\!-\!Ph\!-\!SO_3{}^-H^+$$

Anion-exchanger. Phenyl-bonded silica gel is suspended in chloromethyl methyl ether and boiled under reflux. Then, chloromethyl methyl ether containing anhydrous $ZnCl_2$ is added and the mixture boiled under reflux. After filtration, the gel is washed using 1,4-dioxane and diethyl ether. The washed silica gel is suspended in trimethylamine–1,4-dioxane solution, and the mixture stored at 0 °C to give the quaternary ammonium ion exchanger:

$$\equiv\text{Si–Ph} \rightarrow \equiv\text{Si–Ph–N}^+\text{Me}_3 \text{ Cl}^-$$

Organic Stationary Phase Materials

A variety of organic polymer gels are available for use as stationary phase materials. Their physical strengths and hydrophobicity depend on the properties of the monomers. Dextran gels are polar and soft. Vinyl alcohol copolymer gels are fairly polar. Polyacryl gels are physically hard and relatively non-polar. Polystyrene gels (a copolymer of styrene and divinylbenzene) are physically very strong and very hydrophobic. A higher divinylbenzene concentration makes a physically stronger polystyrene gel. The particle size, pore size, physical strength, and wettability in solvent depends on the reaction solvent, temperature, mixing conditions, and catalyst. Furthermore, ion-exchange groups can be introduced on the polystyrene surface to make ion-exchange resins.

2 Sieving of Stationary Phase Materials

The narrower the particle size distribution, the higher in theory is the potential theoretical plate number. A rough sieving is achieved by a water flow, air flow, or a vibration method. A common sieving method is Hamiltonian water flow (Figure 3.4). The particle distribution can be controlled within ± 1 μm by this method. A slurry of stationary phase material is allowed to float in the cylinder, and a solvent flows from the bottom to the top. The smaller and lighter particles float to the top of the cylinder and the larger and heavier particles sink to the bottom. The required particles are collected at the top of the cylinder. The selection of suspension solvent and control of the temperature are important.

3 Column Packing Methods

Dry packing is usually only used for large particle-size stationary phase materials, especially in preparative-scale columns. High-performance analytical-scale columns are packed by the wet slurry method. Balanced or viscous slurry methods are applied to reduce the sedimentation of the stationary phase materials during the packing process. Previously, balanced density slurries were made by mixing high-density halogenated solvents. However, halogenated solvents are toxic, so the viscous slurry method has been developed, which uses an ethylene glycol–alcohol suspension mixture.

First, an empty analytical column, a pre-column, and a slurry reservoir are connected in series. The narrow-bore analytical column and pre-column are

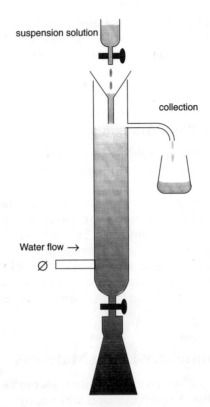

suspension solution

collection

Water flow →

∅

Figure 3.4 *Hamiltonian particle sizing system.*

filled with a packing solvent such as methanol. The reservoir is filled with the stationary phase slurry, and then the packing solvent is quickly pumped through the system under high pressure, depending on the physical strength of the stationary phase material.

For example: for the preparation of a 15 cm long, 4.6 mm i.d. stainless tube column, 2.5 g of octadecyl-bonded silica gel was suspended in 25 ml of hexanol–methanol mixture, and kept in an ultrasonic bath for a few minutes to remove air. After the reservoir was filled with the slurry, methanol was pumped in at 10 ml min^{-1} under constant pressure, 45 MPa (450 bar). After the replacement of slurry solvent by methanol, the flow was stopped and the pressure allowed to drop. When 0 MPa was reached the reservoir was removed. Then, 20 ml of water was added and methanol was again pumped in under the same conditions as before. Again, the flow was stopped and the pressure allowed to drop until it reached 0 MPa. The pre-column was removed and the analytical column closed. The maximum pressure that can be applied in the filling stage is based on the pore size, particle shape, and purity of the silica gel. This reproducible packing procedure is performed at constant temperature by using a water bath (60–80 °C).

4 Column Evaluation

The performance of columns is determined by using a simple procedure for both new and used columns. The test method and the results are usually printed on a sheet provided by the column manufacturer. Each manufacturer uses the most suitable method for their column to obtain the best theoretical plate number. The column evaluation provides important information for all users of liquid chromatography.

Column Efficiency and Asymmetry

Common standard compounds for reversed phase columns are toluene and naphthalene, which have retention factors, k, of about 3. The eluent modifier is methanol or acetonitrile at a concentration of 50–80%, depending on the hydrophobicity of the stationary phase material. For other stationary phase materials, corresponding analytes, with $k = 3$–5, can be used.

The efficiency of a well packed column should be about 95 000 m^{-1} $\pm$ 5% for a 5 μm octadecyl-bonded silica gel column. The equations for the calculation of the number of theoretical plates (N) and peak asymmetry (As) are as follows. A model chromatogram is given in Figure 3.5.

The value of N is given by:

$$N = 5.54 \left(\frac{t_R}{w_{\frac{1}{2}}}\right)^2 = 16\left(\frac{t_R}{w_b}\right)^2 = \left(\frac{t_R}{\sigma}\right)^2$$

where the retention time (t_R) is 5.87 min in Figure 3.5. Assuming the peak is a Gaussian curve, the peak width at the base is considered to be 4σ where σ is the uncertainty in the retention times of individual analyte molecules. The peak width (in the figure) at half-peak-height ($w_{\frac{1}{2}}$) is 0.17 min; and w_b, the peak width at the base line, is 0.30 min. Therefore, N is 6605 from $w_{\frac{1}{2}}$, and 6126 from w_b. The difference is due to the uncertainties in the measurement of the peak width. For comparison between columns of different lengths the efficiency is often expressed as plates per m [$100(N/L)$], where L is the length of the column in cm.

The value of the peak asymmetry As is calculated from:

$$As = b/a$$

where a and b are, respectively, the widths of the front and back of a peak measured at 10% peak height from the baseline. The a and b values (Figure 3.5) are 0.12 and 0.16 min, respectively. Therefore, As is 1.33. The desired As value is 1.0–1.25 for good separations and a long column life. The lifetime of columns with $As < 1.0$ is usually short, due to an inhomogeneously packed bed.

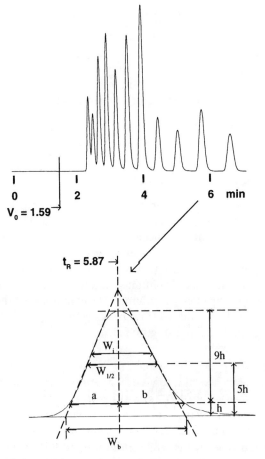

Figure 3.5 *Schematic of a typical chromatogram. V_0, void volume; $w_{\frac{1}{2}}$, width at half peak height; w_b, width at base; 10h, peak height; a, width of front at 1h; b, width of back at 1h; $w_i = 2\sigma.10h$.*

Column Test for Reversed-phase Liquid Chromatography

Inertness Test for Basic Compounds

A bonded stationary phase whose silanol groups are completely modified by alkylation is, in theory, inert towards basic compounds. The lifetime of such columns is long, and can be maintained for more than 2000 h in 1% trifluoro-acetic acid solution (pH 2) or 0.05 M disodium phosphate solution (pH 8.6). Such a column will withstand the passing of more than 10 litres of trifluoro-acetic acid solution for continuous peptide separation. The peak asymmetry of basic compounds is excellent. The extent of unreacted silanol groups can be measured by the chromatographic behaviour of pyridine compared with phenol.[3,4] The relationship of the different surface activity to the number of silanol groups is shown in Figure 3.6. If no active silanol groups are present, the

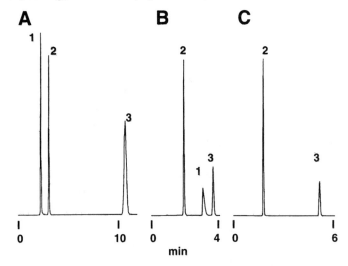

Figure 3.6 *Test for active silanols on octadecyl-bonded silica gels. Column: 5 μm octadecyl-bonded silica gel, 15 cm × 4.6 mm i.d.; eluent, 60% aqueous acetonitrile; flow rate, 1 ml min⁻¹; temperature, ambient; detection, UV 254 nm. Peak 1, pyridine; 2, phenol; and 3, toluene. A, Column with no active silanol groups; B, some active groups; and C, numerous active groups.*

pyridine peak is sharp and its symmetry is like that in chromatogram A. Chromatogram B indicates the existence of some active silanol groups. The pyridine is retained longer than phenol and the peak shape is poor. In chromatogram C pyridine was completely adsorbed and could not be observed. The lifetimes of columns B and C were short in buffered solution. These columns are not suitable for the chromatography of nitrogen-containing basic compounds due to a long tailing peak or the disappearance of the peak by strong adsorption.

Inertness Test for Acidic Compounds

The retention and the peak asymmetry of benzoic acid also indicate the inertness of the bonded phase. If basic compounds remain on the surface or are used as reagents, the peak asymmetry of benzoic acid is poor. The peak height is lower than that of the same quantity of *o*-toluic acid.[3,4] This phenomenon is observed if the basic catalyst that was used in the synthesis process has not been completely washed off the stationary phase or if active amino groups remain. This type of column is not suitable for the separation of acidic compounds.

Inertness Test for Heavy Metals

Active compounds often form a chelate with heavy metal ions, and elution from a column containing heavy metal ions is poor.[3,4] The existence of heavy metals on the surface can be examined by using 8-hydroxyquinoline, which is used for heavy metal extraction from water. However, 8-hydroxyquinoline is regarded as a human carcinogen. Therefore it has to be used under controlled conditions. Other chelation reagents, such as pentan-2,4-dione and 2,3-dihydroxy-naphthalene, can be used for the trace metal test, but they are not as sensitive as 8-hydroxyquinoline. The metal contents of different grades of silica gels are summarized in Table 3.2. Extra pure and pure silica gel contain only trace amounts of heavy metals but contain traces of alkaline metals from the distilled water used mainly during the sizing procedure. The effect of heavy metals in relation to the purity of silica gel is shown in Figure 3.7. The purer the silica gel, the sharper the peak shape, as in chromatogram A. From the peak tailing, Chromatogram B indicates the existence of heavy metals on the surface. The

Table 3.2 *Metal content of silica gels of different purity*

Silica gel	$PS*^a$ (μm)	$ps*^b$ (Å)	$SA*^c$ ($m^2 g^{-1}$)	Metal/ppm						
				Na	Mg	Al	Ca	Ti	Fe	Zn
Ordinary	5	122	333	190	250	150	730	160	22	19
Purified	5	120	350	5	3	120	10	130	50	nd
Pure	10	100	350	6	nd	63	nd	70	13	6
Extra pure	5	100	450	5	1	5	9	nd	1	1

[a] Particle size; [b] pore size; [c] surface area; nd, not detected by inductive coupled plasma spectroscopy (ICP), Cr, Mn, and Ni were not detected.

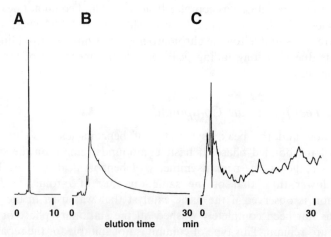

Figure 3.7 *Test for trace metal ions on the surface of octadecyl-bonded silica gel. Column A, 5 μm octadecyl-bonded silica gel, 15 cm $\times$ 4.6 mm i.d.; B, octadecyl-bonded silica gel, 15 cm $\times$ 4.6 mm i.d.; C, octadecyl-bonded 10 μm silica gel, 30 cm $\times$ 4.0 mm i.d.; eluent, 50% aqueous acetonitrile; flow rate, 1 ml min^{-1}; temperature, ambient; detection, UV 220 nm (the scales of A, B, and C are not the same); sample, 8-hydroxyquinoline. Columns; A, metal ion free; B, low proportion of metal ions; C, high level of metal ions.*

peak of 8-hydroxyquinoline could not be observed on chromatogram C. This also indicates that the surface treatment was poor. Pure silica gels (purity > 99.99%) are necessary for the separation of porphyrins. For the separation of biologically active compounds, bonded phases made from pure silica gels whose purity is known and guaranteed should be used.

Physicochemical Tests of the Stationary Phase

The carbon content of a stationary phase is measured by an elemental analyser, as a weight balance before and after heating at 800 °C. Particle size, pore size, and surface area are measured by specific instruments, such as a particle size analyser, nitrogen adsorption porosimeter, and mercury depression analyser, respectively. The precision of the measurement of carbon content is high; however, that of the other measurements is relatively poor. Therefore, it is difficult to relate the surface area of different silica gels to analyte retention factors.

5 Measurement of Void Volume

The retention factor, k, is the basic value in chromatography, and is related to the void volume (dead volume). The void volume is the space inside the column, where no retention of solutes has occurred and can be measured on a chromatogram, as shown in Figure 1.3. The void volume is about half the total volume of the column when it is packed with porous stationary phase materials. In practice, the effective void experienced by the analyte is smaller because the molecular mass of the analyte is usually much greater than that of the eluent molecule. In a model of porous stationary phase material, the pores can be represented as V-shape valleys (Figure 3.8), where region a is a support, such as

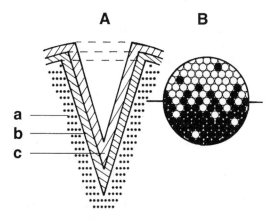

Figure 3.8 A, *Schematic structure of porous stationary phase*; a, *support*; b, *bonded phase*; and c, *adsorbed solvent layer*. B, *detailed view of eluent components in surface layer*; ●, *lower phase molecules*; ○, *upper phase molecules*.

silica gel or polystyrene gel, b is the bonded phase, such as octadecyl or ion-exchange groups, and c is the solvent layer, which is the adsorbed components of the eluent. In Figure 3.8, if a sample molecule is small it can pass to the bottom of the valley. However, a larger sample molecule, such as a protein, is large, and 'jumps' the valley if the pore size is small. The real void volume for a particular analyte thus depends upon its shape and mass. The pore size distribution of the stationary phase materials also affects the void volume of individual analytes. The exact value cannot be measured. A practical compromise method is examined below.

In normal-phase liquid chromatography, the elution volume of the solvent is nearly constant in eluents having different mixtures of solvent components. The solvent peak can therefore be used to measure the void volume. In reversed-phase liquid chromatography, the thickness of the organic solvent layer, which is c in Figure 3.8A, depends on the concentration of organic modifier and additives. In addition, an ion-exclusion effect is observed for ionized solutes, and the elution volume is often less than the exclusion limit of the column. However, measurement of the void volume is necessary to calculate k. Figure 3.8B indicates the microscopic interface of a partition. The ● symbols are lower-phase molecules and ○ are upper-phase molecules in a solvent partition. The distribution between the phases b and c is not clear-cut, as in a partition model. Phase b incorporates the bonded phase and the concentration of the organic component of the eluent is higher than in the bulk eluent; it gradually reduces towards the mobile phase in reversed-phase liquid chromatography. The effective void volume is therefore not constant.

Elution Volume of Proposed Void Volume Markers

Several compounds have been proposed for the measurement of the void volume, including sodium nitrate solution, water, deuterium oxide, fructose, acetonitrile, tetrahydrofuran (THF), meso-erythritol, gluconolactone, and 2,4-dinitronaphthol. The elution volume of a number of these compounds has been measured in 10–90% aqueous acetonitrile and acidic–aqueous acetonitrile. The results are given in Figure 3.9 where the volumes in A and B were measured in aqueous acetonitrile and in C and D were measured in aqueous acetonitrile containing 50 mM phosphoric acid. Methanol (a) and deuterium oxide (g) showed two peaks when monitored by a refractive index detector (Figure 3.9C).

Even the elution volume of acetonitrile was not constant in these eluents and varied from 1.25 to 1.95 ml. The volumes of ionized sodium nitrate (e) and 2,4-dinitronaphthol (d) were smaller than the exclusion limit (1.0 ml) of this column. The dramatic change for 2,4-dinitronaphthol in Figure 3.9A indicates that the volume depended strongly upon the ratio of eluent components.[5]

In acidic eluents (Figure 3.9C and D), the retention of acidic compounds becomes stronger and that of basic compounds becomes weaker. In this system, uric acid (h) is not suitable as a void volume marker due to its longer retention time. In neutral and basic eluents, an ionized acid can be used as the marker because no other compounds are eluted more rapidly. Fructose (c) is a very

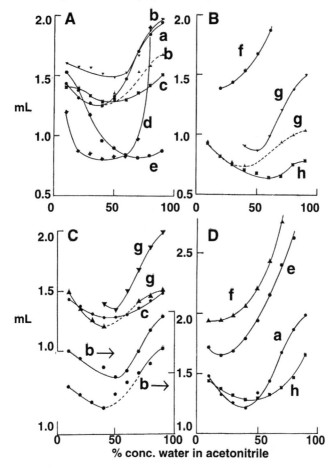

Figure 3.9 *Elution volumes (ml) of alternative void volume markers. Column, 5 μm octadecyl-bonded silica gel, 15 cm × 4.5 mm i.d.; eluents A and B, 10–90% aqueous acetonitrile, eluents C and D, 10–90% aqueous acetonitrile containing 50 mM phosphoric acid; flow rate, 1 ml min⁻¹; temperature, 30 °C; detection, UV 210 nm and refractometer. Sample: a, acetonitrile; b, methanol; c, fructose; d, 2,4-dinitronaphthol; e, sodium nitrate; f, tetrahydrofuran; g, deuterium oxide; and h, uric acid.*

polar compound, and is not adsorbed on a hydrophobic stationary phase surface. It is also UV-active at short wavelengths and non-ionizable under practical conditions. Its elution volume is nearly constant in the above conditions. The effect of injection volume and the necessity of a refractive index detector are thus avoidable. Therefore, fructose can be used as the standard void volume measuring marker. When analytes such as steroids are examined, they are excluded from many of the pores and the molecular size of fructose is too small to measure their void volume. Maltotriose may then be the compound of choice for the void volume measurement.

6   Selectivity of Stationary Phase Materials

Silica Gel-based Stationary Phase Materials

Separations can be performed on a single column by using different eluents. Practical separations can also be achieved using column selection. In over 50% of applications octadecyl-bonded silica gel is used as the hydrophobic stationary phase material. As an alternative, phenyl-bonded silica gel is less hydrophobic compared with octadecyl- or octyl-bonded silica gel. This means that a smaller amount of organic modifier is necessary to obtain the same retentions. In addition, aromatic compounds are relatively more highly retained on phenyl-bonded silica gel than on octadecyl-bonded silica gel, and alkanes are more retained on octadecyl-bonded silica gel than on phenyl-bonded silica gel as shown in Figure 3.10. Such selectivity can be useful for the separation of compounds having different degrees of aromaticity.

Many manufacturers sell the same types of stationary phase materials,[6] but the most popular stationary phase materials, *e.g.* octadecyl-bonded silica gels, from different sources, even from the same manufacturer, often demonstrate different retention capacities and selectivities. Such differences are due to the aggressive reactivity of the silylation method and the different bonding reactions that are used, as described earlier.

The selectivity differences can be understood by the comparison of four octadecyl-bonded silica (ODS) phases synthesized from the same silica gel. The

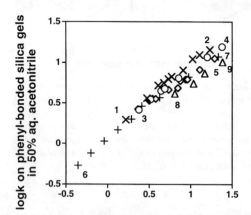

**logk on octadecyl-bonded silica gels
in 70% aq. acetonitrile**

Figure 3.10   *Selectivity of phenyl-bonded silica gel compared with ODS silica gel.
Columns: ODS-silica, YMC ODS, 15 cm × 6.0 mm i.d., phenyl-bonded
silica gel, YMC phenyl, 10 cm × 6.0 mm i.d.; eluent, aqueous acetonitrile.
Compounds: ×, polycyclic aromatic hydrocarbons; ○, alkylbenzenes;
◇, polychlorobenzenes; +, alkanols; △, alkanes; 1, benzene; 2, benzopyrene;
3, toluene; 4, heptylbenzene; 5, hexachlorobenzene; 6, hexanol; 7, tetra-
decanol; 8, pentane; and 9, octane.*

four materials are a high-carbon (HIC)-loaded silica gel with and without a second silylation (end-capping) (HIC-ODS-E and HIC-ODS-NE, respectively) whose carbon content is 16 wt%, and a low-carbon (LOC)-loaded silica gel with and without a second silylation (LOC-ODS-E and LOC-ODS-NE, respectively) whose carbon content is 8.9 wt%.[7] The selectivity-related hydrophobicity differences can be compared using the differences in the retention factors of homologous alkylbenzenes as the van der Waals volume increases (Figure 3.11). The hydrophobicity, *i.e.* retention capacity, of HIC-ODS was found to be about twice that of LOC-ODS. The end-capping treatment further increased the retention capacity. This result indicates that if the quality control of the bonding processes is poor, the products will have different retention capacities. This difference also depends upon the surface area of silica gels.

The relationship of the selectivity to the polarity of the analytes can be understood from the differences in the retention factors of homologous alkanols (Figure 3.12). The polar alkanols are relatively more retained on the non-endcapped bonded phases (LOC-ODS-NE and HIC-ODS-NE) because smaller-size alkanols can reach the unreacted silanol groups on the surface of silica gels.

The relationship of the selectivity towards π-electrons can be understood from the differences in the retention factors of polycyclic aromatic hydrocarbons (Figure 3.13). The difference in the retention factors on end-capped and non-endcapped stationary phase materials is less than that of alkylbenzenes. This is due to the water content of the stationary phase. The content may be higher in non-endcapped bonded phases.

The best octyl- and octadecyl-bonded silica gels should be chemically stable in highly basic solutions. The well-bonded phases give good results in the inertness

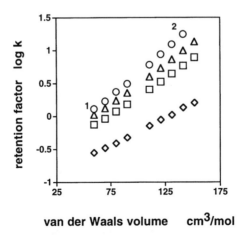

Figure 3.11 *Comparison of the retention of toluene to decylbenzene as a measure of hydrophobicity of different octadecyl-bonded silica gels with different carbon loadings. Columns:* □, *LOC-ODS-E;* ◇, *LOC-ODS-NE;* ○, *HIC-ODS-E;* △, *HIC-ODS-NE; eluent, 80% aqueous acetonitrile at 30 °C. Compounds: 1, toluene; 2, nonylbenzene.*

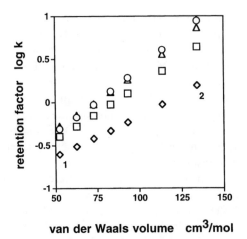

Figure 3.12   *Polar selectivity of different octadecyl-bonded silica gels towards the reten-*
tion of homologous alkanols. Columns: □, LOC-ODS-E; ◇, LOC-ODS-
NE; ○, HIC-ODS-E; △, HIC-ODS-NE; eluent, 70% aqueous acetonitrile
at 30°C. Compounds: 1, butanol; 2, dodecanol.

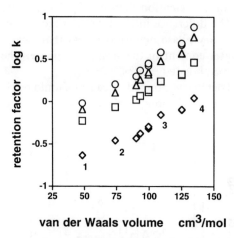

Figure 3.13   *π-Electron selectivity of different octadecyl-bonded silica gels towards the*
retention of polycyclic aromatic hydrocarbons. Columns: □, LOC-ODS-E;
◇, LOC-ODS-NE; ○, HIC-ODS-E, △, HIC-ODS-NE; eluent, 80%
aqueous acetonitrile at 30°C. Compounds: polycyclic aromatic hydro-
carbons; 1, benzene; 2, naphthalene; 3, pyrene; 4, 3,4-benzopyrene.

tests (see above), and can be used for continuous analysis in basic solutions, pH
9.0, over a one month period. Such stability and inertness are guaranteed by the
manufacturers. All silica-loaded materials are unstable in high pH eluents. The
amount of carbon loading can be more than 110% of the theoretical calculation; however, the accuracy of surface area measurement is less than that of
the carbon loading measurement. Non-endcapped bonded phases sometimes

demonstrate better relative separations; however, their long-term operation is not guaranteed. The lifetime of short-chain bonded phases, such as butyl, cyano, and amino, is also short in buffer solutions due to their weak hydrophobicity. It seems that these phases do not reject the adsorption of ions from the eluent on to the untreated silanol groups.

Organic Polymer-based Stationary Phase Materials

Surface-modified silica gels are used for a variety of separations, but organic polymer-based stationary phase materials are more useful for long-term operations, such as for an amino acid analyser and size-exclusion liquid chromatography.

Physically strong organic porous and non-porous polymer gels are also used for high-performance liquid chromatography. When difficulty in achieving separation is encountered with silica gel-based stationary phase materials, organic gels will often solve this problem even though the theoretical plate number may be less than that of silica gel-based stationary phase materials.

For example, polystyrene gel is very hydrophobic, and aromatic compounds are selectively retained. Polystyrene oligomers are separated on a polystyrene gel using isocratic elution (Figure 3.14). Vitamins, anions, and nucleic acids and bases have been chromatographed on polystyrene gel.[8] These separations can also be carried out on alkyl group-bonded silica gels, if the silanol groups are completely covered.[3] Large pore-size polystyrene gels are useful for the separation of proteins. Proteins with $M_r = 6000–66\,300$ have been separated on a

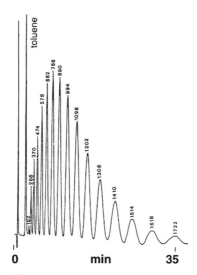

Figure 3.14 *Reversed-phase liquid chromatogram of polystyrene oligomers. Conditions: columns, PLRP-S, 15 cm × 4.6 mm i.d., 100 Å (polystyrene gel); eluent, 70% aqueous tetrahydrofuran; flow rate, 1 ml min^{-1}; detection, UV 254 nm. Numbers beside peaks indicate the molecular weight.*
(Reproduced by permission from Polymer Laboratories data)

300 Å pore size polystyrene gel, and $M_r = 120\,000$ collagen and $340\,000$ fibrinogen have been separated on a 1000 Å pore-size polystyrene gel, as shown in Figure 3.15. Silica gel-based bonded stationary phase materials with a 1000 Å pore size are not as physically strong as the narrow pore-size silica gels. However, polystyrene gels have the advantages of surface inertness and moderate physical strength.

Methacrylate polymers are as physically strong as polystyrene gels. Their hydrophobicity is weaker than that of polystyrene gel, and the aromatic selectivity is similar to that of octadecyl-bonded silica gel. A chromatogram of hydantoins on a methacrylate polymer is shown in Figure 3.16. The elution order is related to the hydrophobicity of the chemicals.

Vinyl alcohol copolymer gel is hydrophilic and has been developed for aqueous-phase size-exclusion liquid chromatography; however, it is less polar than the polysaccharides. Its specificity permits the direct injection of a biological sample without deproteinization. For example, blood serum from a patient suffering from chronic nephritis has been injected directly as a measure of the degree of dialysis (Figure 3.17). Adenosine triphosphate, adenosine diphosphate, and adenosine monophosphate in red blood cells have also been separated directly (Figure 3.18). Theophylline in blood serum has been

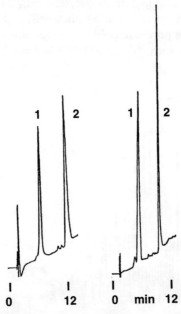

Figure 3.15 *Chromatogram of fibre-type proteins on polystyrene gels having different pore sizes. Column A, PLRP-S 300 Å, 15 cm × 4.6 mm i.d.; B, PLRP-S 1000 Å (polystyrene gel), 15 cm × 4.6 mm i.d.; eluent, 15 min linear gradient from 20% of 0.25% trifluoroacetic acid to 60% of 0.25% trifluoroacetic acid in 95% aqueous acetonitrile; flow rate, 1.0 ml min^{-1}; detection, UV 220 nm. Peaks: 1, collagen (M_r 120 000) and 2, fibrinogen (M_r 340 000).* (Reproduced by permission from Polymer Laboratories data)

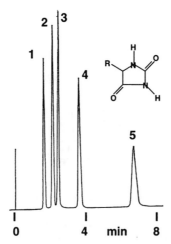

Figure 3.16 *Chromatogram of hydantoins on polymethacrylate stationary phase. Conditions: columns, Shodex RSPak DE613 (methacrylate gel); eluent, 1:1 mixture of 0.033 M disodium hydrogen phosphate and potassium dihydrogen phosphate; flow rate, 2.0 ml min^{-1}; detection, UV 210 nm; temperature, 50°C. Compounds: 1, R = (CH$_2$)$_2$CO$_2$H; 2, R = CH$_2$OH; 3, R = H; 4, R = CH$_3$; and 5, R = C$_2$H$_5$.*
(Reproduced by permission from Shodex data)

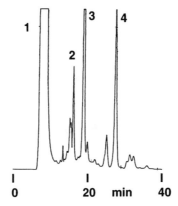

Figure 3.17 *Chromatogram of blood serum from a patient with chronic nephritis. Conditions: columns, Asahipak GS320 (vinyl alcohol copolymer gel), 50 cm × 7.6 mm i.d.; eluent, 0.1 M sodium phosphate containing 0.3 M sodium chloride pH 7.0; flow rate, 1 ml min^{-1}; detection, UV 250 nm; direct injection of sample. Peaks: 1, protein; 2, orotidine; 3, creatinine; and 4, uric acid.*
(Reproduced by permission from Asahipak data)

separated from proteins (Figure 3.19). In these separations, the proteins were eluted first and the targeted compounds eluted later; the quantitative analysis of the target compounds is thus easy. Pesticides in blood serum have also been directly analysed on vinyl alcohol copolymer gels, as shown in Figure 3.20.

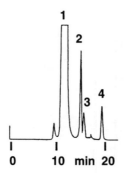

Figure 3.18   *Adenosine phosphates in blood on vinyl polymer column. Conditions: column, Asahipak GS320 (vinyl alcohol copolymer gel), 50 cm × 7.6 mm i.d.; eluent, 0.1 M sodium phosphate buffer containing 3 M sodium chloride pH 7.0; flow rate, 1.0 ml min⁻¹; detection, UV 260 nm. Peaks: 1, haemoglobin; 2, adenosine triphosphate; 3, adenosine diphosphate; and 4, adenosine monophosphate.*

(Reproduced by permission from Asahipak data)

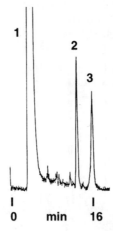

Figure 3.19   *Theophylline in blood serum on vinyl alcohol polymer column. Conditions: column, Asahipak GS320 (vinyl alcohol copolymer gel), 50 cm × 7.6 mm i.d.; eluent, 0.01 M sodium acetate buffer pH 4.0 in 10% aqueous acetonitrile; flow rate, 2 ml min⁻¹; detection, UV 280 nm. Peaks: 1, protein; 2, low M_r impurity; and 3, theophylline.*

(Reproduced by permission from Asahipak data)

These separations can be carried out using a silica-based bonded phase; however, the important advantage of organic polymer stationary phase materials is their chemical stability. The columns can be washed by using an alkaline solution after a certain number of injections. According to the chromatograms, the proteins in serum are completely eluted and nothing remains inside the column. However, the pressure drop in this type of analysis

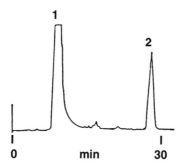

Figure 3.20   *Pesticide in rat blood serum on vinyl alcohol polymer column. Conditions:*
column, Asahipak GS320 (vinyl alcohol copolymer gel), 50 cm × 7.6 mm
i.d.; eluent, 0.05 M phosphate buffer containing 0.2 M sodium chloride pH 7.2
in 20% aqueous methanol; flow rate, 1.0 ml min⁻¹; detection, UV 254 nm.
Peaks: 1, blood serum protein and 2, methomyl.
(Reproduced by permission from Asahipak data)

may increase after several injections, depending on the injection volume. This
problem is not avoidable on all types of columns. Therefore, washing using an
alkaline solution is occasionally required.

The separation of basic and metal-sensitive compounds is difficult on silica-
based stationary phase materials, but these separations can be performed on
vinyl alcohol copolymer gels. Examples are the separation of methallothionein
from dolphin kidney, α-, β-, and γ-endorphin, and nucleotide and nucleoside
mixtures.[8] However, an analytical-scale separation may also be performed on
surface-modified wide-pore silica gels (pore size 300 Å or more), using columns
which showed a negative response in the heavy metal test described above.

Modified vinyl alcohol copolymer gels with phenyl groups improved the
hydrophobicity and selectivity for aromatic compounds. Chromatograms of
protein mixtures and enzymes (phosphoisomerase) are shown in Figures 3.21
and 3.22, respectively. The separation mechanism on the surface of these
stationary phase materials is basically the same as that for reversed-phase
liquid chromatography, and is called hydrophobic liquid chromatography. The
chromatograms of protein mixtures shown in Figure 3.23 were obtained on a
phenyl-bonded vinyl alcohol copolymer gel and a trimethylsilyl-bonded silica
gel using basically the same eluent. The elution order of proteins was the same,
but the resolution was a little different. Octadecyl-bonded vinyl alcohol
copolymer gels are chemically stable and overcome the weakness of octadecyl-
bonded silica gels, which are influenced by unreacted silanol groups and high
pH instability.[9]

The selection of inorganic or organic polymer stationary phase materials
depends on the application. When information is needed rapidly, silica-based
bonded phases have the advantage due to their physical strength and high
theoretical plate number, although their lifetime is shorter. When a longer-term
operation is necessary, such as for an amino acid analyser, organic polymer-
based stationary phase materials are appropriate. Organic polymer-based

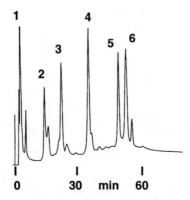

Figure 3.21   *Separation of proteins on wide pore phenyl-bonded column. Conditions:*
column, TSKgel Phenyl-5 PW, 75 mm × 7.5 mm i.d.; eluent, 60 min linear
gradient of ammonium sulfate concentration from 1.7 to 0 M in 0.1 M
phosphate buffer (pH 7.0); flow rate, 1.0 ml min^{-1}; detection, UV 280 nm.
Peaks: 1, cytochrome c; 2, myoglobin; 3, ribonuclease; 4, lysozyme; 5, α-
chymotrypsinogen; and 6, α-chymotrypsin.
(Reproduced by permission from Tosoh data)

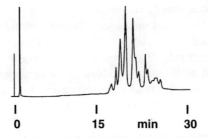

Figure 3.22   *Chromatogram of phosphoisomerase. Conditions: column, TSKgel Phenyl-5*
PW RP, 75 mm × 4.6 mm i.d.; eluent, 2 min linear gradient of acetonitrile
from 0.05 to 20% followed by 48 min linear gradient of acetonitrile from 20
to 80% in 0.05% trifluoroacetic acid; flow rate, 1 ml min^{-1}; detection, UV
220 nm.
(Reproduced by permission from Tosoh data)

stationary phase materials can be recycled, but silica-based stationary phase
materials are used only once; this is another selection point, which depends on
the economic considerations.

The disposal of used stationary phase materials, especially those used for
biomedical applications, should also be of concern. Silica gel-based stationary
phase materials can be heated at high temperature before disposal, and organic
polymer-based stationary phase materials can be burned. Take care.

Other Stationary Phase Materials

Hydroxyapatite and titania are also useful stationary phase materials for
biological samples. Hydroxyapatite, $Ca_{10}(PO_4)_6(OH)_2$, as a hexagonal column

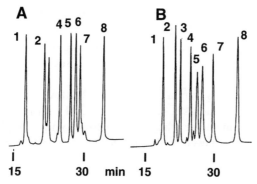

Figure 3.23 *Selectivity of phenyl and alkyl bonded stationary phase materials for protein separation. Column A, TSK gel phenyl-5PW RP, 75 mm × 4.6 mm i.d.; B, TSK gel TMS 250, 75 mm × 4.6 mm i.d.; eluent, 60 min linear gradient elution from 5% of 0.05% trifluoroacetic acid in 5% aqueous acetonitrile to 80% of 0.05% trifluoroacetic acid in 80% aqueous acetonitrile; flow rate, 1 ml min^{-1}; detection, UV 220 nm. Peaks: 1, ribonuclease; 2, insulin; 3, cytochrome c; 4, lysozyme; 5, transferrin; 6, bovine serum albumin; 7, myoglobin; and 8, ovalbumin.*
(Reproduced by permission from Tosoh data)

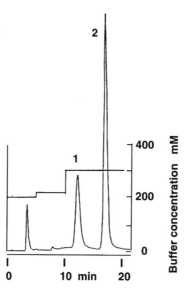

Figure 3.24 *Separation of plasmid DNA on a titania column. Conditions: column, HCA column A-7610 (titania), 10 cm × 7.6 mm i.d.; eluent, step-wise gradient of potassium phosphate buffer concentration (pH 6.8); flow rate, 1.0 ml min^{-1}; detection, UV 260 nm. Peaks: 1, purasmid DNA (pBR 322) and 2, cromorimal.*
(Reproduced by permission from Polymer Laboratories data)

crystal is formed as a porous stationary phase material. The large pore size, 700–10 000 Å, is suitable for purification of bio-polymers. Bovine serum albumin, lactic dehydrogenase, ribonucleic acid, and immunoglobulin in blood serum can be purified. Hydroxyapatite is also used as a support for affinity liquid chromatography.

The pore size of porous titania can be up to 2000 Å. Titania is used for the purification of proteins and as a support for bound enzymes. The purification of β-lactoglobulin from cheese whey, of protease from pineapple, β-lactamase, and amylase can be achieved with titania. The latter two purifications are impossible on alumina. Titania is also used as a support in peptide synthesis. The separation of plasmid DNA is shown in Figure 3.24.

7   References

1. G.B. Alexander (ed.), 'Silica and I', Doubleday & Company, 1967.
2. K.K. Unger (ed.), 'Porous Silica', Journal of Chromatography Library, Vol. 16, Elsevier, Amsterdam, 1979.
3. M. Ohhira, F. Ohmura, and T. Hanai, *J. Liq. Chromatogr.*,1989, **12**, 1065.
4. T. Hanai, M. Ohhira, and T. Tamura, *LC-GC*, 1988, **6**, 922.
5. L. Boksanyi, O. Liardon, and E. sz Kovats, *Adv. Colloid Interface Sci.*, 1976, **6**, 95.
6. T. Hanai, in 'Encyclopedia of Analytical Science', Academic Press, London, 1995, pp. 2558–2567.
7. J. Yamaguchi, T. Hanai, and Hong Cai, *J. Chromatogr.*, 1988, **441**, 183.
8. Asahipak data from Showa Denko.
9. K. Yasukawa, Y. Tamura, T. Uchida, Y. Yanagihara, and K. Noguchi, *J. Chromatogr.*, 1987, **410**, 129.

CHAPTER 4

Selection of the Eluent

The most important parameter for the control of liquid chromatography is the composition of the eluent. Liquid chromatography is a powerful separation method with unlimited possibilities of eluent selection. However, it is not easy to choose a suitable eluent within a short time without a number of trial experiments. The crucial factor is to control the solubility of the analytes in the eluent. Increasing the solubility of analytes in the eluent decreases their retention times. The selection of the components of an eluent is described below, based on the properties of the analytes to be separated. The important properties are hydrophobicity, dipole moment, hydrogen bonding, ionization, and steric effects.

1 Reversed-phase Liquid Chromatography

The most popular system, with over 50% of separations, is reversed-phase liquid chromatography on octadecyl-bonded silica gel. Analytes are retained on the stationary phase materials based on their hydrophobicity. This means that polar compounds are eluted faster than non-polar compounds. The higher the hydrophobicity of the stationary phase surface, the longer will be the retention times of the analytes. Bonding of donor-electron groups onto the stationary phase surface results in a stronger retention of compounds containing dipoles. In some cases, interactions between untreated silanol groups on the stationary phase and hydroxy groups on the analyte increase the selectivity; however, polar groups on the stationary phase are generally eliminated to obtain inert stationary phase materials, which are more stable in long-term operation.

Selectivity of Organic Modifiers in the Eluent

Three organic solvents, acetonitrile, methanol, and THF, are usually used as the organic modifiers. Increasing the concentration of the organic modifier decreases the overall retention times, but changes in relative retention times depend on the properties of the analytes.

The general relationship between the type of solute and its retention can be seen by comparing the retention factors, k, of a set of standard compounds with their octanol–water partition coefficients, *i.e.* the $\log P$ value (listed in Table 4.1), as a measure of their relative solubility in water. The logarithm of the retention factor, $\log k$, of these compounds measured in 50% aqueous acetonitrile on an octadecyl-bonded silica gel column shows a close linear relationship (Figure 4.1).

However, the elution behaviour of aromatic and aliphatic compounds is often different (even though their carbon numbers and van der Waals volumes are very similar) in eluents containing different organic modifiers. These variations are due to differences in the solubility of analytes in the organic solvent.

Five organic solvents [acetonitrile, methanol, tetrahydrofuran (THF), acetone, and dimethylformamide], which are homogeneously miscible with water, have been used as modifiers to study the relationship of the selectivity of the solvent to the molecular properties of analytes. The polar interaction

Table 4.1 *Log P values of standard analytes*

No.	Chemical	Log P	No.	Chemical	Log P
1	Pentanol	1.33	11	Benzene	2.28
2	Hexanol	1.86	12	Naphthalene	3.21
3	Heptanol	2.39	13	Fluorene	3.91
4	Octanol	2.92	14	Anthracene	4.38
5	Decanol	3.98	15	Pyrene	5.03
6	Dodecanol	5.04	16	Chrysene	5.79
7	Methyl benzoate	2.15	17	Toluene	2.59
8	Isopropyl benzoate	3.09	18	Ethylbenzene	3.12
9	Butyl benzoate	3.74	19	Isopropylbenzene	3.52
10	Isopentyl benzoate	4.14			

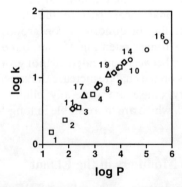

Figure 4.1 *Retention factors related to log P values. Column, 5μm octadecyl-bonded silica gel (LiChrosorb LC7) 25 cm × 4.1 mm i.d.; eluent, 50% aqueous acetonitrile; flow rate, 1 ml min⁻¹. Compounds: □, alkanols; ◇, benzoates; ○, polycyclic aromatic hydrocarbons; and △, alkylbenzenes.*

Table 4.2 *Interaction properties of solvents for reversed-phase liquid chromatography*

Solvent	Po	Xe	Xd	Xn
Tetrahydrofuran	4.0	1.52	0.80	1.68
Acetone	5.1	1.78	1.17	2.14
Methanol	5.1	2.45	1.12	1.58
Acetonitrile	5.8	1.80	1.57	2.44
Dimethylformamide	6.4	2.50	1.34	2.56
Water	10.2	3.77	3.77	2.55

Po: polarity, *Xe*: proton acceptor, *Xd*: proton donor, *Xn*; dipole moment.

properties of the solvents were calculated from reference values,[1] and are listed in Table 4.2. The values of the polarity (*Po*) and the polar interaction properties, proton acceptor (*Xe*), proton donor (*Xd*), and dipole moment (*Xn*), of the mixed solvents can be calculated from their concentrations.

As above, the $\log k$ values of the different types of compounds [alkanols (ROH), polycyclic aromatic hydrocarbons (PAH), alkyl benzenes (RB), and alkyl benzoates (ROB)] each demonstrated a linear relationship with their $\log P$ values and this relationship was observed with different ratios of acetonitrile and water. The experiment was then performed in eluents containing different organic modifiers and it was found that the behaviour of analytes containing different functional groups differed and the effect depended on the $\log P$ of the analyte.

To demonstrate these differences, the experimental relationship between $\log P$ and $\log k$ across a range of eluent compositions was determined for each group of analytes. The results were then used to calculate the predicted $\log k$ in each case for a theoretical model compound with $\log P = 5$ as a hydrophobic model, for $\log P = 1$ as a hydrophilic model, and for $\log P = 3$ as an intermediate model.

The results in a range of aqueous acetonitrile eluents are shown in Figure 4.2. For both alkanols and polycyclic aromatic hydrocarbons the predicted values for each of the model compounds fall on a single line represented by the solid points (●) (as expected from Figure 4.1). However, with THF as the modifier, predicted values for the model alkanols (ROH) (◆) and polycyclic aromatic hydrocarbons (PAH) (○) differ and the relative positions change with the $\log P$ values. The $\log k$ values of $\log P = 5$ alcohols and PAHs in the aqueous THF eluent were relatively smaller than those obtained for the alkanols and PAHs in aqueous 30–70% acetonitrile, but the $\log k$ values of the $\log P = 1$ model compounds were higher. Furthermore, in aqueous THF the $\log k$ values of polycyclic hydrocarbons were larger than those of alkanols. These results indicate that when a separation between aromatic and polar aliphatic compounds is necessary, THF may be an effective organic modifier.

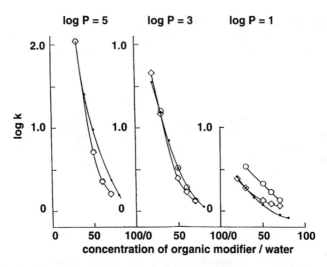

Figure 4.2 *Predicted chromatographic behaviour in aqueous acetonitrile and tetrahydro-furan of model compounds whose log P values are 1, 3, and 5. Column, 5 μm octadecyl-bonded silica gel (LiChrosorb LC-7), 25 cm × 4.1 mm i.d. Compounds and eluent: •, polycyclic aromatic hydrocarbons and alkanols in aqueous acetonitrile; ◇, alkanols in aqueous tetrahydrofuran; and ○, poly-cyclic aromatic hydrocarbons in aqueous tetrahydrofuran.*

Comparison of the Effect of Organic Modifiers on Analytes with log P = 3

The selectivity of a number of organic modifiers was examined using the predicted $\log k$ values of the $\log P = 3$ models from each group in different organic modifier–water mixtures. The composition of the eluent was adjusted so that either the solubility parameter,[1] polarity (*Po*), proton acceptor (*Xa*), proton donor (*Xd*), or dipole moment (*Xn*) values were kept constant to determine which parameter affected the selectivity. The results are summarized in Table 4.3.

The compositions of the different eluents were determined using the values in Table 4.2. The composition was adjusted in each case so that the selected polarity term was equal in each eluent. For example, in Table 4.3 the polarity (p_0) of the 40:60 THF–H_2O is $(0.4 × 4) + (0.6 × 10.2) = 7.72$, when *Xe*, 2.87; *Xd*, 2.58; *Xn*, 2.20. This has an equivalent polarity to 56:44 ACN–H_2O $[(0.56 × 5.8) + (0.44 × 10.2) = 7.72$ but now *Xe*, 2.66; *Xd*, 2.53; *Xn*, 2.49].

Chromatographic Behaviour of log P = 3 Compounds in Aqueous Tetrahydro-furan. For analytes whose $\log P = 3$, in 40% aqueous THF the predicted $\log k$ values of the model alkanols (ROH), polycyclic hydrocarbons (PAH), alkyl-benzenes (RB), and alkyl benzoates (ROB) are 0.64, 0.77, 0.90, and 0.79, respectively. The polarity of 40% aqueous THF is 7.72. The same polarity can be obtained in 56% aqueous acetonitrile. As noted earlier, in aqueous aceto-

Table 4.3 *Solubility problems of eluents*

| Eluent | Po | Xe | Xd | Xn | Predicted log k of log P = 3 compound | | | |
					ROH	PAH	RB	ROB
40% THF–H$_2$O	**7.72**	**2.87**	**2.58**	**2.20**	0.64	0.77	0.90	0.79
56% ACN–H$_2$O	**7.72**	2.66	2.53	2.49			0.40	
46% ACN–H$_2$O	8.19	**2.87**	2.76	2.50			0.63	
54% ACN–H$_2$O	7.82	2.70	**2.58**	2.49			0.46	
nc	nc	nc	nc	**2.20**			nc	
% of ACN–H$_2$O to give the same log k as 40% THF					46	40	36	43
60% MeOH–H$_2$O	**7.14**	**2.98**	**2.18**	**1.97**	0.81	0.63	0.67	0.62
69% ACN–H$_2$O	**7.14**	2.40	2.24	2.47			0.18	
40% ACN–H$_2$O	8.44	**2.98**	2.89	2.50			0.78	
72% ACN–HO	7.02	2.35	**2.18**	2.47			0.14	
nc	nc	nc	nc	**1.97**			nc	
% of ACN–H$_2$O to give the same log k as 60% MeOH					39	46	44	46
50% ATN–H$_2$O	**7.65**	**2.78**	**2.47**	**2.35**	0.35	0.33	0.41	0.31
58% ACN–H$_2$O	**7.65**	2.63	2.50	2.49			0.37	
50% ACN–H$_2$O	7.99	**2.78**	2.66	2.49			0.53	
59% ACN–H$_2$O	7.60	2.61	**2.47**	2.48			0.36	
% of ACN–H$_2$O to give the same log k as 50% ATN					59	61	56	62
50% DMF–H$_2$O	**8.30**	**3.13**	**2.56**	**2.55**	1.26	0.69	0.87	0.89
43% ACN–H$_2$O	**8.30**	2.92	2.82	2.50			0.70	
33% ACN–H$_2$O	8.77	**3.13**	3.06	2.51			1.02	
55% ACN–H$_2$O	7.78	2.69	**2.56**	2.49			0.43	
0% ACN–H$_2$O	10.2	3.77	3.77	**2.55**			nc	
% of ACN–H$_2$O to give the same log k as 50% DMF					26	44	37	36

Po: Polarity, *Xe*: proton acceptor, *Xd*: proton donor, *Xn*: dipole moment, THF: tetrahydrofuran, ACN: acetonitrile, MeOH: methanol, ATN: acetone, DMF: dimethylformamide, nc: cannot be calculated.

nitrile the log k values for all compounds with log $P = 3$ are the same and thus the predicted value for the ROH, PAH, RB, and ROB models is 0.40, unlike any of the THF retentions. These results indicate that chromatographic behaviour cannot be controlled only by polarity. The proton acceptor value (*Xe*) of the 40% aqueous THF eluent was 2.87. The same proton acceptor value can be obtained in 46% aqueous acetonitrile (Table 4.3), where the predicted log k values of each of the log $P = 3$ model compounds is 0.63. Thus in this case the value of the log k of ROH is nearly equal to the value in 40% aqueous THF (0.64).

If the proton donor value (*Xd*) is fixed at 2.58, the equivalent eluent is 54% aqueous acetonitrile and the predicted log k value for the log $P = 3$ compounds should be 0.46, which is again different from any of the log k values in 40% aqueous THF. The dipole moment (*Xn*) of 40% aqueous THF was 2.20; however, such a value cannot be obtained with any proportion of aqueous acetonitrile. To obtain the same log k values as in 40% aqueous THF, a composition of 46% aqueous acetonitrile is needed for ROH, 40% aqueous acetonitrile for PAH, 36% aqueous acetonitrile for RB and 43% aqueous acetonitrile for ROB. This comparison demonstrates the different selectivity of THF.

Chromatographic Behaviour of log P = 3 Compounds in Aqueous Methanol. In a similar calculation, the $\log k$ values of ROH, PAH, RB, and ROB in 60% aqueous methanol are 0.81, 0.63, 0.67, and 0.62, respectively (Table 4.3). The same $\log k$ values can be obtained in 39% aqueous acetonitrile for ROH, in 46% for PAH, 44% for RB and in 46% for ROB. The values of *Po*, *Xe*, and *Xd* (but not *Xn*) in 60% aqueous methanol can be obtained in 69% for *Po*, in 40% for *Xe*, and in 72% aqueous acetonitrile for *Xd*. Only the $\log k$ of ROH in 40% aqueous acetonitrile demonstrated a nearly equal $\log k$ value to that obtained in 60% aqueous methanol. Thus the solvent properties do not have a specific relationship with the $\log k$ values of these compounds.

Chromatographic Behaviour of log P = 3 Compounds in Aqueous Acetone. When the *Xe* value in aqueous methanol was nearly equal to that in aqueous acetonitrile (see above), the $\log k$ of ROH was about the same as that in aqueous acetonitrile but the corresponding predicted value with aqueous acetone (log $k = 0.35$) is very different to the equivalent aqueous acetonitrile (log $k = 0.53$) (Table 4.3). However, when the *Po* and *Xd* values in aqueous acetonitrile are nearly equal to those in aqueous acetone, the predicted $\log k$ values of OH are about the same. To obtain the same $\log k$ as in 50% aqueous acetone it was necessary to use 59% aqueous acetonitrile for ROH, 61% for PAH, 56% for RB and 62% for ROB.

Chromatographic Behaviour of log P = 3 Compounds in Aqueous Dimethylformamide (DMF). DMF is a solvent with a strong dipole moment (Table 4.2). The same $\log k$ values in 50% aqueous DMF can be obtained at a concentration of 26% aqueous acetonitrile for ROH, 44% aqueous acetonitrile for PAH, 37% for RB, and 36% for ROB. When eluents with equivalent *Po*, *Xe*, *Xd*, or *Xn* values in 50% DMF were selected, different $\log k$ values were predicted in all three cases.

   The above results only demonstrate the chromatographic behaviour of compounds with $\log P = 3$. Their relative retention factors thus are influenced by the properties of solvents in the eluent. The relative values cannot be controlled by only one property of the solvents and chromatographic behaviour therefore also depends on the properties of the analytes.

Comparison of Effects of Modifiers on a Mixture of Analytes

The following results demonstrate the comparative chromatographic behaviour of the 19 compounds listed in Table 4.2. In reversed-phase liquid chromatography, changing the concentration of organic modifier affects both the overall retention time and the relative elution behaviour of different types of compounds. The effect of the organic modifier especially influences the retention of polycyclic aromatic hydrocarbons, as seen in Figure 4.3A (compounds 11–16). The results in 50% aqueous acetone and aqueous acetonitrile indicate that the retention times in 50% aqueous acetone are shorter than those in 50% aqueous acetonitrile. A selectivity difference for alkanols (●, 1–6) is not observed but the retention time of the larger PAHs is shorter than the values expected from their size.

From the results in 40% aqueous THF and 50% aqueous acetonitrile (Figure 4.3B), the relative retention of PAHs (11–16) is weak and that of alkylbenzenes (○, 17–19) is strong in aqueous THF. The alkyl group affects the strong retention in aqueous THF. As found earlier (Table 4.3), a methanol–water

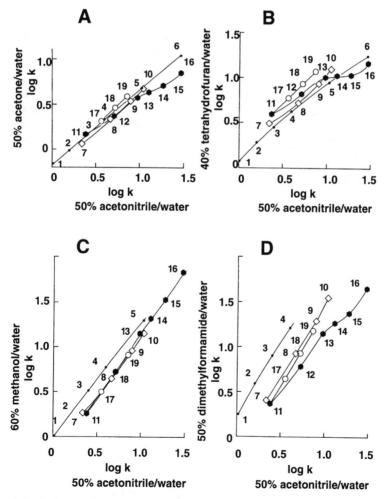

Figure 4.3 *Relative selectivity of different organic modifiers. Numbers beside symbols, 1–6 (●) ROH; 7–10 (◇) ROB; 11–16 (filled hexagons) PAH; and 17–19 (○) RB.*

mixture (Figure 4.3C) can distinguish alkanols (•, ROH) from the other analyte groups.

Comparison of the retention behaviour in 50% aqueous DMF and acetonitrile (Figure 4.3D) indicated that DMF is a selective modifier for the separation of alkanols (•, 1–6) and PAHs (11–16) and of aromatic and aliphatic compounds. Unfortunately, DMF is too aggressive as a solvent and attacks organic polymer piston seals in the pump and, therefore, highly concentrated solutions cannot be used as an eluent.

From the theoretical viewpoint, acetonitrile is the most suitable solvent to study the correlation of retention times and log P values of analytes, since the dipole moment (2.44) is nearly equal to that of water (2.55) (Figure 4.4). The electron donor effect can therefore be eliminated, and the elution order is not changed on modification of the acetonitrile–water mixture ratio. The first choice of an eluent should therefore be an acetonitrile–water mixture for non-ionic compounds in reversed-phase liquid chromatography. Methanol, acetone, THF, or DMF can then be added to improve the resolution.

Example 1: Separation of an alkylbenzene and an alkyl benzoate. The log P values of butyl benzoate and isopropylbenzene are 3.74 and 3.52, respectively, from Table 4.1. The separation of these compounds in 50% aqueous acetonitrile therefore requires a high plate number column as the log k values will be nearly identical. The most suitable eluent is 40% aqueous THF. See compounds 9 and 19 in Figure 4.3B.

Example 2: Separation of a polycyclic aromatic hydrocarbon and an alkyl benzoate. The log P values of fluorene and butyl benzoate are 3.91 and 3.74, respectively, from Table 4.1. The separation is poor in 50% aqueous aceto-

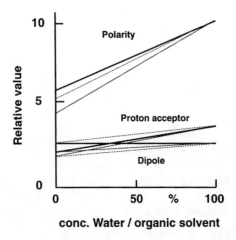

Figure 4.4 *Polarity, proton acceptor and dipole properties of water and organic modifier mixtures. ——, acetonitrile; – – –, methanol; and ···· tetrahydrofuran.*

nitrile; however, they can be separated in 60% aqueous methanol or 50% aqueous DMF (compounds 13 and 9, Figures 4.3C and 4.3D).

However, in selecting a mobile phase the entire analytical system and separation system should be considered. In many cases, a component of a good eluent may not be suitable for research purposes. For example, the use of water as an eluent in methods that will require further chemical reactions and purification should be avoided due to the difficulty in concentrating the products of interest after the separation and the stability of the products in the process. In addition, organic solvents are flammable, combustible, and toxic. The threshold limit concentrations of acetone, acetonitrile, DMF, methanol, and THF are 250, 20, 10, 200, and 200 ppm, respectively. The most easily handled solvent of these five is methanol. Acetone is the least used solvent due to the strong UV absorption (UV cut-off = 330 nm). THF is unstable and easily oxidized by oxygen in air, causing a cut-off at longer wavelengths as it ages.

At present, the selection of an organic modifier is estimated from the aliphatic or aromatic nature of analytes. However, the properties of analytes often cannot be easily obtained. Examples of quantitative structure–retention relationships based on the $\log P$ and van der Waals volume of analytes are demonstrated in Chapter 6.

2 Chromatography of Ionic Compounds

Neutral compounds are retained by hydrophobicity in reversed-phase liquid chromatography, and by hydrogen bonding and/or π–π interactions on hydrophilic phases in normal-phase liquid chromatography. Ionization of the analyte reduces the retention time in reversed-phase liquid chromatography, but in some cases increases the retention time on ion-exchangers. Retentions on ion-exchange resins made from polystyrene gel are based both on ion–ion interactions with ion-exchange groups and on hydrophobic interactions with the polystyrene gel matrix. The addition of counter-ions can improve the resolution.

pH Effects

The retention factor of partially ionized compounds can be predicted in reversed-phase liquid chromatography by Equation (4.1):[2]

$$k = \frac{k_0 + k_i(K_a/[H^+])}{1 + (K_a/[H^+])} \tag{4.1}$$

where k_0 is the maximum retention factor of the unionized form of the analyte. The term k_i is the retention factor of the fully ionized compound; K_a is the dissociation ionization constant; and $[H^+]$ is the hydrogen ion concentration of the eluent. Every compound has its own k_0, k_i, and K_a values. When the hydrophobicities of compounds are nearly equal, separation is difficult in the reversed-phase mode. However, when their dissociation constants are different, the separation can be easily accomplished in a pH-controlled eluent by introducing differential partial ionization. The value of k_i is close to zero in a reversed-phase system; however, the value is not negligible in the ion-exchange mode.

Example: separation of 3-chloro and 4-ethylbenzoic acids. The log P values of these compounds are similar, and their pK_a values are 3.96 and 4.35, respectively. As seen by their chromatographic behaviour in Figure 4.5, the separation is difficult at low pH due to the similar log P values giving similar log k_0 values. The separation can be easily performed in an eluent with a pH of about 4.2, due to the difference in pK_a values and different degrees of ionization.

The retention factor of an amino acid is given by the following equation, as the amino acid has two ionizable groups (a carboxyl group and an amino group, with K_{a1} and K_{a2}, respectively):

$$k = \frac{k_0 + k_1([H^+]/K_{a1}) + k_{-1}(K_{a2}/[H^+])}{1 + [H^+]/K_{a1} + K_{a2}/[H^+]}$$

The maximum retention factor (k_0) is related to the log P value and k_1 and k_{-1} are the retention factors of the cationic and anionic forms, respectively. The pK_a values are known, and the retention factor in a given eluent can therefore be predicted in reversed-phase liquid chromatography using an alkyl-bonded silica gel or polystyrene gel column. The separation conditions can be adjusted according to their log P and pK_a values by the selection of a suitable organic modifier concentration and the pH of the eluent.[3,4]

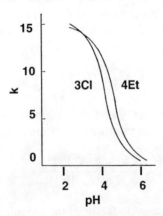

Figure 4.5 *Effect of pH on the retention factors of 3-chloro- and 4-ethylbenzoic acids. Column, octadecyl-bonded vinyl alcohol copolymer gel, 10 cm × 6 mm i.d.; eluent, 0.05 M sodium phosphate solution in 20% acetonitrile; flow rate: 1 ml min^{-1}; temperature, 30 °C. Compounds: 3Cl, 3-chlorobenzoic acid; and 4Et, 4-ethylbenzoic acid.*

The separation of compounds whose pK_a values are different can be carried out easily on an ion-exchanger using ion–ion interactions.[5] For example, non-ionized mandelic acid is eluted before benzoic acid due to weak hydrophobicity. But mandelic acid is retained more than benzoic acid on an ion-exchanger due to its stronger acidity. With polystyrene gel-based ion-exchangers both ion–ion and hydrophobic interactions are important factors in the retention mechanism. Compounds whose pK_a values are similar can be separated on an ion-exchange resin. For example, amino acids have been separated by cation-exchange liquid chromatography (Figure 4.6) and carboxylic acids by anion-exchange liquid chromatography (Figure 4.7). The elution order of these separations was directly related to their dissociation constants. The high temperature of the separation was also important, as were the acidic and ionic strength of the eluents.

The separation of very hydrophobic compounds, such as proteins, is difficult on an ion-exchange resin with a hydrophobic matrix, because organic modifiers are required as part of the eluent. However, the separation can be performed on an ion-exchanger with a hydrophilic matrix. A weak acidic ion-exchanger is suitable for the separation of basic proteins, and a weak basic ion-exchanger is suitable for the separation of acidic proteins (Figures 4.8 and 4.9, respectively). The components of the eluents should not be harsh, to prevent the denaturation of the proteins. A salting-out effect can be used to obtain a gradient elution. A combination of buffer solutions and weak ion-exchangers on a polar matrix, such as poly(vinyl alcohol), is suitable for the separation of bio-polymers, especially for preparative-scale separations. Oligonucleotides from polyadenyl acid and RNA of *Escherichia coli* have been separated on weak anion-exchangers using ammonium acetate buffer and a sodium chloride concentra-

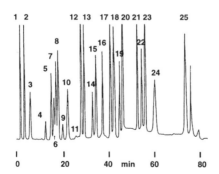

Figure 4.6 *Separation of amino acids in body fluid. Column, MCI GEL CK10V, 15 cm × 4.6 mm i.d.; eluent, four-step gradient of sodium citrate buffer; flow rate, 0.5 ml min⁻¹; temperature, 60 °C; detection, fluorescence reaction detector of sodium hydrochlorite/orthophthaldehyde, Ex. 360 nm, Em. 450 nm. Peaks: 1, SSA; 2, TAU; 3, GCGA; 4, HPro; 5, Asp; 6, Gln + Asp; 7, Thr; 8, Ser; 9, Sar; 10, Glu; 11, Pro; 12, Gly; 13, Ala; 14, Cys; 15, Val; 16, Met; 17, Ile; 18, Leu; 19, Tyr; 20, Phe; 21, His; 22, Orn; 23, Lys; 24, Trp; and 25, Arg.*
(Reproduced by permission from Mitsubishikasei data)

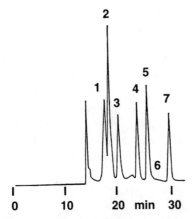

Figure 4.7 *Anion exchange separation of carboxylic acids in red wine. Column, Shodex C811, 100 cm × 7.6 mm i.d.; eluent, 3 mM perchloric acid; flow rate, 0.9 ml min⁻¹; temperature, 60 °C; detection, reaction detection using chlorophenol red at 430 nm. Peaks: 1, citric acid; 2, tartaric acid; 3, malic acid; 4, succinic acid; 5, lactic acid; 6, formic acid; and 7, acetic acid.*
(Reproduced by permission from Shodex data)

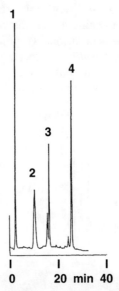

Figure 4.8 *Cation-exchange liquid chromatography of basic proteins. Column, Asahipak ES502C; eluent, 20 min linear gradient of sodium chloride from 0 to 500 mM in 50 mM sodium phosphate buffer pH 7.0; flow rate, 1 ml min⁻¹; temperature, 30 °C; detection, UV 280 nm. Peaks: 1, myoglobin from horse skeletal muscle (M_r 17 500, pI 6.8–7.3); 2, ribonuclease from bovine pancreas (M_r 13 700, pI 9.5–9.6); 3, α-chymotrypsinogen A from bovine pancreas (M_r 257 000, pI 9.5); and 4, lysozyme from egg white (M_r 14 300, pI 11.0–11.4).*
(Reproduced by permission from Asahikasei data)

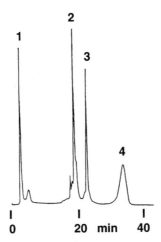

Figure 4.9 *Anion-exchange liquid chromatography of acidic proteins. Column, Asahipak ES502N; eluent, 20 min linear gradient of sodium chloride from 0 to 500 µM in 50 mM bis-tris/HCl buffer pH 7.0; flow rate, 1 ml min⁻¹; detection, UV 280 nm. Peaks: 1, conalbumin (Mr 77 000–88 000, pI 6.0–6.8); 2, ovalbumin (Mr 45 000, pI 4.6); 3, trypsin inhibitor (Mr 8000, pI 4.5); and 4, β-lactoglobulin (Mr 18 400, pI 5.1).*
(Reproduced by permission from Asahikasei data)

tion gradient.[6,7] Subfragments of myosin have been separated on a weak cation exchanger using a sodium chloride concentration gradient.[8] The selection of eluent components such as pH and salt concentration makes possible the direct analysis of catecholamines in urine without deproteinization and of haemoglobin in red blood by cation-exchange liquid chromatography, as shown in Figures 4.10 and 4.11, respectively.

Figure 4.10 *Direct analysis of catecholamines in urine sample. Column, Asahipak ES-502C; eluent, 75 mM succinic acid + 25 mM borate buffer (pH 6.10) containing 0.5 mM EDTA; flow rate, 1.0 min⁻¹; detection, fluorescence reaction detection Ex. 350 nm. Peaks: 1, adrenaline; 2, noradrenaline; and 3, dopamine.*
(Reproduced by permission from Asahikasei data)

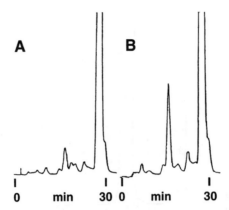

Figure 4.11   *Direct injection analysis of haemoglobin in red blood cells. Column, Asahipak ES-502C; eluent, 32 min linear gradient from 25% 30 mM sodium phosphate buffer to 65% 30 mM sodium phosphate containing 300 mM sodium chloride pH 5.5; flow rate, 1.0 ml min⁻¹; detection, 425 nm. Samples: A, normal subject and B, diabetic patient.*

Ion-pair Separations

Ionic compounds are usually separated by ion-exchange liquid chromatography, and non-ionic compounds are separated by normal-phase or reversed-phase liquid chromatography. The separations of mixtures of ionic and non-ionic compounds and of different ionic and polar compounds are more difficult. One method for extracting such compounds from aqueous solution is ion-pair extraction where the ionic charge on the analyte is neutralized by the addition of a counter-ion of opposite charge, and the analyte is extracted into an organic phase as the neutral ion-pair. This ion-pair extraction method was first applied as ion-pair partition liquid chromatography by Schill *et al.*[9]

The selection of the counter-ion and its concentration are important for the separation of ionic compounds in reversed-phase and ion-exchange liquid chromatography. The addition of hydrophobic ions is an especially powerful method and several surfactants can be used as hydrophobic counter-ions. The theoretical column efficiency of ion-pair liquid chromatography is much better than that of an ion-exchange column, and the regeneration of a column is much faster. Thus, if we can control ion-pair liquid chromatography, we can solve a separation problem. (The important background sources in this area are listed at the end of the chapter.)

Theory of Ion-pair Liquid Chromatography

An ionized solute A^- forms a paired ion AB with a counter-ion B^+ in water. The paired ion is extracted into an organic phase from the aqueous phase, as given by the equation:

$$A_{aq}^- + B_{aq}^+ \rightarrow AB_{org}$$

where aq is the aqueous phase and org is the organic phase. When applying this ion-pair extraction, the following parameters must be considered.

a. Selection of stationary phase material.
b. Selection of counter-ion and the concentration.
c. pH and ionic strength of eluent.
d. Selection of organic modifier and the concentration.
e. Selection of column temperature.
f. Selection of detector.

Two types of system are used for ion-pair liquid chromatography. When polar stationary phase materials, such as silica gel, are used an ion-pair partition mechanism is applied. When non-polar stationary phase materials, such as octadecyl-bonded silica gel and polystyrene gel, are employed a paired-ion adsorption mechanism is involved. The former is called normal-phase ion-pair partition liquid chromatography, and the latter is called reversed-phase ion-pair liquid chromatography.

The following conditions should be satisfied to perform satisfactory ion-pair liquid chromatography.

a. A one-to-one ion-pair must be formed.
b. The equilibrium concentration of the ion-pair in the aqueous phase must be very low compared with that of the ion.
c. Only the paired-ion is extracted into the organic phase.
d. The counter-ion makes an ion-pair only with the target ion.
e. The paired-ion is not decomposed in the eluent.
f. The partition equilibrium is related only to the chromatographic retention.

Normal-phase Ion-pair Partition Liquid Chromatography

Analytes in the organic phase (eluent) form ion-pairs with counter-ions existing in the hydrophilic stationary phase. The ion-pairs are then partitioned into the organic phase. Usually, porous silica gel is used as the support, and water or formamide is the stationary phase. The requirements are that a specific amount of counter-ions must be contained in the stationary phase, and the pH must be controlled to ionize the analytes. The eluent is not miscible with water, and counter-ions must stay in the stationary phase.

The conditions were employed for the separation of sulfa drugs[10] using tetrabutylammonium hydrogen sulfate (TBAHSO$_4$). The pH of this counter-ion is greater than 6 in water; it exists as tetrabutylammonium cation (TBA$^+$) and sulfate anion (SO$_4^{2-}$). n-Pentane was used as the major component of the eluent with a small amount of n-butanol, which enhanced the solubility of the paired-ion. A small amount of water was added to the above mixture, which was then left standing until equilibrium was reached. A proportion of the paired-ion moved into the organic phase and a small amount of butanol and n-pentane moved into the aqueous phase. In this equilibrated system, the pH in the stationary phase is constant before and after equilibration. In the aqueous phase the amount of *n*-butanol is *ca.* 4–5% (w/w) and that of *n*-pentane is negligible. The amount of water in the organic phase depends on the concentration of *n*-butanol. In 25% *n*-butanol–*n*-pentane (v/v), the water concentration in the organic phase is 1.5% (w/v), and 5.0% in the case of 50% (v/v) *n*-butanol–*n*-pentane. These values meet the above requirements; however, it is not simple in practice to obtain the appropriate experimental conditions.

Reversed-phase Ion-pair Liquid Chromatography

In this mode analytes form hydrophobic ion-pairs in the eluent, and the paired-ions are retained on a hydrophobic stationary phase surface by their hydrophobicity. The hydrophobic paired ions are then eluted by reversed-phase liquid chromatography. Gradient methods, such as organic modifier gradients, counter-ion gradients, ionic-strength gradients, and pH gradients, can be applied in this system. Usually less than 0.1 wt% of the counter-ion and buffer concentration is employed. The selection of a stationary phase material is simple and a variety of hydrophobic stationary phase materials for reversed-phase liquid chromatography can be used. The organic modifier concentration depends on the hydrophobicity of stationary phase materials used and the ion-pair reagent. Increasing the size of the counter-ion increases the retention time. The maximum retention time is reached when the counter-ion concentration reaches the micelle condition. The relationship between counter-ions and analytes is summarized in Table 4.4.

Example: ion-pair liquid chromatography of amino acids. Amino acids are zwitterions. The amino group can form an ion-pair with an alkanesulfonate ion (such as octanesulfonate), and the carboxyl group can form an ion-pair with a tetrabutylammonium ion, depending on the pH of the solution.

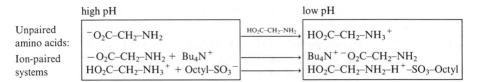

In reversed-phase liquid chromatography, the ionization of the solute decreases the retention. The addition of counter-ion under these conditions forces the formation of an ion-pair between the ionized solute and counter-ion, and then the retention of the analyte increases as the paired-ion is retained.

Aspartic and glutamic acids were not retained in a buffer solution without a counter-ion; however, these acids were retained by the addition of octyl sulfate to the eluent, as seen in Figure 4.12. These amino acids can form a complex with copper ions and will be retained on the stationary phase. The addition of both a counter-ion and copper ions further increased the retention and made possible

Table 4.4 *Relationship between analytes and counter-ions in ion-pair liquid chromatography*

Analytes	Counter-ions

Strong and weak acids
  sulfate dyes
  carboxylic acids
  sulfa drugs
	Quaternary amines
	  tetrabutyl ammonium
	  tetramethyl ammonium
	  hexadecyl trimethylammonium
	  ammonia
	Tertiary amines
	  trioctylamine

Strong and weak bases
  catecholamines
	Alkyl-, allyl-sulfonic acids
	  pentane, hexane, heptane, octane, decane,
	  dodecane sulfonic acids
	Phosphoric and sulfonic acids
	Alkyl sulfate, dodecyl sulfate[a]

Basic compounds	Perchloric acid

[a] Use as alkyl sulfonic acids, the selectivity is slightly different.

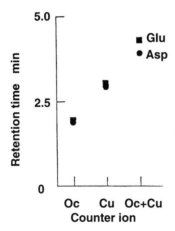

Figure 4.12 *Effect of counter-ions and copper on the retention of amino acids. Column, octadecyl-bonded silica gel, 25 cm × 4.6 mm i.d.; eluent, 0.01 M sodium acetate buffer (pH 5.6) containing 1.2 mM sodium octanesulfonate (Oc) and/ or 0.1 mM copper acetate (Cu); flow rate, 1 ml min^{-1}; detection, UV 220 nm. Compounds: Glu, glutamic acid; Asp, aspartic acid.*

the separation of these compounds. Increasing the alkyl chain length of the counter-ion also enhanced the retention, as seen in Figure 4.13.

Increasing the concentration of the counter-ion further increased the retention, but the retention reached a plateau, as seen in Figure 4.14. A total separation of the amino acids by reversed-phase ion-pair liquid chromatography could be performed.[11] A column switching technique reduced the total analysis time (Figure 4.15).

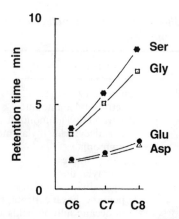

Figure 4.13   *Effect of alkyl chain length of counter-ion on separation of amino acids. Column, octadecyl-bonded silica gel, 15 cm × 4.6 mm i.d.; eluent, 0.01 M sodium acetate buffer containing 0.4 M copper acetate and 1.0 mM sodium alkanesulfonate (pH 5.6); flow rate, 1 ml min*[−1]*; detection, UV 230 nm. Counter-ion: C6, sodium hexanesulfonate; C7, sodium heptanesulfonate; and C8, sodium octanesulfonate. Compounds: Ser, serine; Gly, glycine; Glu, glutamic acid; and Asp, aspartic acid.*

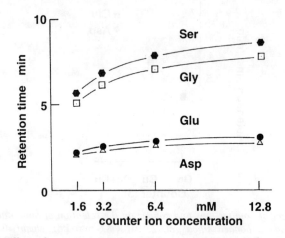

Figure 4.14   *Effect of counter-ion concentration on separation of amino acids. Column, octadecyl-bonded silica gel, 15 cm × 4.6 mm i.d.; eluent, 0.01 M sodium acetate buffer (pH 5.6) containing 0.4 mM copper acetate and 1.6–12.8 mM sodium heptanesulfonate; flow rate, detection, and analytes, see Figure 4.13.*

Two mechanisms for retention in reversed-phase ion-pair liquid chromatography have been considered. One is the adsorption of the hydrophobic paired ion on the hydrophobic surface of stationary phase material. In the second mechanism, the hydrophobic counter-ion is held on the surface of the hydrophobic stationary phase, and the analyte ion is retained by ion–ion interactions, as shown in Figure 4.16. In the latter case, of a dynamic ion-exchange

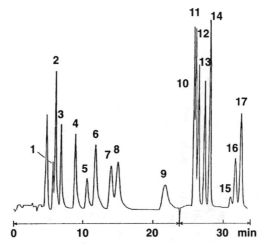

Figure 4.15 *Ion-pair liquid chromatography of free amino acids using a column switching system. Column 1, butyl-bonded silica gel, 50 × 4.6 mm i.d., 2, octyl-bonded silica gel, 50 × 4.6 mm i.d., and 3, octadecyl-bonded silica gel, 250 × 4.6 mm i.d.; eluent, 0.01 M sodium acetate buffer (pH 5.6) containing 4 mM copper acetate and 0.8 mM sodium heptanesulfonate; flow rate, 1 ml min⁻¹; detection, UV 235 nm. Peaks: 1, Tyr; 2, Val; 3, Met; 4, His; 5, Lys; 6, Ile, 7, Leu; 8, Phe; 9, Arg; 10, Asp; 11, Ser; 12, Glu; 13, Thr; 14, Gly; 15, Pro; 16, Cys; and 17, Ala. 1–9 were separated on column 1 and 10–17 were separated by a combination of columns 2 and 3.*

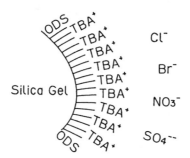

Figure 4.16 *Model of reversed-phase ion-pair liquid chromatography in which counter-ions are held on the stationary-phase surface.*

mechanism, the column efficiency is much higher than in the corresponding ion-exchange liquid chromatography using an ion-exchange resin as the stationary phase. Examples of samples, counter-ions, and stationary phases are summarized in Table 4.5.

pH Effects

The retention of acids in reversed-phase liquid chromatography can be predicted by Equation 4.1. The same equation can be applied to predict the

Table 4.5 *Applications of reversed-phase ion pair liquid chromatography*

Sample	Counter-ion	Stationary phase material	Eluent	Ref.
Alkaloid	TMA	ODS silica	50 mM TMA in 20% MeOH, pH 6.1	24
Alkanesulfonic acid	Cetylpyridium	ODS silica	60% aq. MeOH	17
Tris(2,2'-dipyridyl)ruthenium(II)mono & dicarboxylic acids	Alkylsulfonic acid	ODS silica	0.01 M Borate buffer pH 10.0	18
Amine	Tetrabromophenolphthalein	NH_2 or OH silica	Dichloromethane/MeOH	45
Amino acid	Naphthalene-2-sulfo-dodecanesulfonic acid	Phenyl silica	50 mM Phosphoric acid	14
	Dodecyl sulfate	C8 silica	20% ACN + 50 mM H_2SO_4, 5 mM $NaHSO_4$	15
	TFA	Me silica	Na citrate + Na sulfate + PrOH + SSD	62
Aminoglycoside	PFPa + TFA	C18 silica	0.2 M TFA	38
	Bis(2-ethylhexyl)phosphate	Polystyrene	PFPa + TFA	39
Aminophenol	Trimethylnonylammonium	Silicone/cellulose	Citric buffer pH 3.8	12
Antidepressive amine	Dodecyl sulfate	C8 silica	NaBr + trimethylnonyl ammonium/Pentanol, phosphate buffer	57
Apomorphine	TBA	Ph silica	20 mM Na_2HPO_4 + 30 mM citric buffer pH 3.25 + MeOH + ACN (55:36:9)	35
Anions	Tridecylammonium	C18 silica	TBA + salicylic acid/pH 4.62	47,55
Ascorbic acid	TBA	C18 silica	Tridecylammonium, pH 5.0	56
Barbital	TBA	Butylnitrile/C2 silica	0.03 M TBA pH 7.7	19
Benzoates	TBA	Silanized silica	PeOH + phosphate buffer	65
Benzoic acid derivatives	1-Phenetyl-2-pyrro-liniumcetride	C2 silica	0.03 M TBA pH 7.4	19
Carboxylic acid	Octyl sulfate	Phenyl silica	Acetate buffer pH 4.6	14
	TBA	C8 silica	35% ACN + 0.05 M ammonium acetate	15
Catecholamine		C18 silica	50 mM Citric acid + 50 μM EDTA in 50 mM Na_2HPO_4	34
Cefaclor/cefradine		C18 silica	Phosphate buffer pH 7.0 + TBA + ACN (2:1)	41,54

Choline/3,5-dinitro-benzoic acid derivatives	Dodecyl sulfate	C18 silica	5 mM Dodecyl sulfate, 0.1% acetic acid in 50% ACN	23
Cocaine	Ammonium	C18 silica	70% MeOH + 50 mM NH_4CO_2H	16
Cosmetic dyes	$NaHCl_4$	C8 silica	$NaClO_4$/ACN	48
Daunorubicin	Dodecyl sulfate	C18 silica	Dodecyl sulfate/phosphoric acid + ACN	50
Diazonium salts	Heptanesulfonate	C18 silica	Heptanesulfonate/phosphate buffer ACN	60
L-3,4-Dihydroxyphenylalanine metabolites	Heptanesulfonate	C18 silica	HSA + MeOH + acetic acid	66
Dipeptide	Naphthalene-2-sulfonate	Ph silica	50 mM Phosphoric acid	14
Dyes	TBA	C18 silica	5 mM TBA/45% MeOH	20
Enkephalins	Octanesulfonate	C18 silica	Octanesulfonate + citric acid/phosphate buffer, MeOH	59
Erythromycin	Ammonium	C18 silica	70% MeOH + NH_4OH pH 4.8	16
Guanidino compounds	Octanesulfonate	C8 silica	Octanesulfonate + ninhydrin + ACN + MeOH	37
Heroin	TBA	PrNH silica	5 mM TBA + ACN (15:85)	33
Indoleacetic acid	TBA	Pentanol/C2 silica	0.03 M TBA pH 7.4	19
Iodine, Thiosulfate	TBA	C18 silica	TBA/MeOH + phosphate buffer	49
Neostigmine, pyridostigmin, edrohonium	Pentane sulfate		Pentanesulfonate + NaH_2PO_4/ACN pH 3	21
Oligodeoxyribonucleotides	TBA	C18 silica	2 mM TBAP + ACN/Phosphate buffer pH 7.0/gradient	36
Organic acids	Iron(II) 1,10-phenanthroline	Polystyrene	$Fe(phen)_3SO_4$/0.5 mM acetate buffer	46
Oxymorphone	Octanesulfonate	C8 silica	3 mM Phosphate buffer pH 4 + octanesulfonate/ACN	42
Penicillamine	Heptanesulfonate	C18 silica	1 g Heptanesulfonate + 150 mg EDTA/1 l water pH 4.0	32
Peptide	Ammonia, TFA	C18 silica	50 mM NH_4CO_2H/50 mM TFA, pH 1.3	27
Phenylacetic acid	TBA	Silanized silica	Phosphate buffer + PeOH	63
Phenylenediamine	Ammonium	C18 silica	30% MeOH + 20 mM NH_4OH pH 6.4	16
Pirlimycin	TBAP	C8 silica	TBAP/ACN + perchloric acid	43
Prostaglandin E1	TBA	C18 silica	TBA + MeOH + water + acetic acid (0.15:62:35:3)/gradient	21

(continued)

Table 4.5 Continued

Sample	Counter-ion	Stationary phase material	Eluent	Ref.
Pyrimidine, Purine, Azapurine metabolite	TBA	C18 silica	10 mM TBA, pH 8.0 in 0.2% ACN	22
Pyrocarbine, Phyzostygmin	Pentanesulfonate	C18 silica	5 mM Heptanesulfonate, pH 3.6 in 4% MeOH	29
Quaternary amine	1-Phenetyl-2-pyrrolinium	Phenyl silica	Acetate buffer pH 4.6	14
Rebamisol	Heptanesulfonate	C18 silica	2% Heptanesulfonate in 0.2% acetic acid + MeOH (55 + 4)	30
Riboxamide	HTAB	C18 silica	1 mM HTAB in 10 mM phosphate buffer pH 6.0	28
Salbutamol	Heptanesulfonate	C8 silica	5 mM Heptanesulfonate in 5.8 mM phosphate buffer pH 6.0 + MeOH (31:69)	31
Sulfonamide	TBA	Butylnitrile/C2 silica	0.03 M TBA pH 7.9	19
Sulfonic acids	Cetyltrimethylammonium	SAS silica	PrOH/H_2O or PrOH + CH_2Cl_2/H_2O	64
Sulfonates	TBA	Silanized silica	PeOH + phosphate buffer	65
Tetracycline	$HClO_4$	SC-TAS	ACN	61
Tartrazine	TBA	C18 silica	TBA/formic acid + NaOH	59
Tetrahydrozoline	Octanesulfonate	C18 silica	Octanesulfonate + N,N-dimethyloctylamine/MeOH	51
Theophylline metabolites	TBA	C18 silica	TBA, pH 4.75 in 50% MeOH, gradient	26
	TBA	C18 silica	TBA + 15 mM Tris buffer + ACN + MeOH	40
Tricyclic antidepressant	Decylamine	C18 silica	Decylamine/ACN	53
	Alkylammonium	C8, C18 silica	MeOH/phosphate buffer pH 2	44
Tris(2,2'-bipyridyl)Ru	Heptanesulfonate	C18 silica	5 mM Heptanesulfonate buffer pH 3.5	18
N,N-Trimethylene-bis (pyridium-4-aldoxime) dibromide	Heptanesulfonate	C18 silica	Heptanesulfonate/ACN	58

retention factor in reversed-phase ion-pair liquid chromatography. The pH and counter-ion effects on the chromatographic behaviour of indoleacetic acid are demonstrated in Figure 4.17.

The addition of a surfactant counter-ion reduced the retention factor at low pH due to the surface modification of the stationary phase material. Covering the surface of the stationary phase with the surfactant reduces the hydrophobicity of the stationary phase material. The addition of the tetrabutylammonium counter-ion increased the retention factor at high pH. The pK_a of the indole acetate was 5.15 without surfactant, 4.85 with octyl sulfate ion, and 5.60 with tetrabutylammonium ion. That is, the addition of a same-charged hydrophobic ion reduced the pK_a value, and the addition of the counter-ion increased the pK_a value. The difference in the pK_a value on the addition of surfactant is not constant; it is affected by the kind of ion and the concentration. It is difficult to estimate the pK_a change.

The retention factor of the molecular form of the analyte, k_m, and the retention factor of the paired-ion, k_{ip}, can be easily measured experimentally. When the retention factor is measured at low and high pH, and the pK_a value is known, k at a given pH is roughly estimated from the following equation:

$$k = \frac{k_m + k_{ip}K_a/[H^+]}{1 + K_a/[H^+]}$$

where K_a is the dissociation constant of the ion-pair complex.

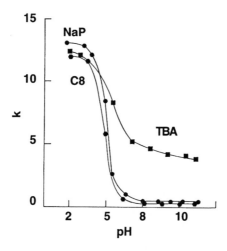

Figure 4.17 *Selectivity of counter-ion for the retention of indoleacetic acid. Column, Hitachi 3011 (polystyrene gel), 25 cm × 4.1 mm i.d.; eluent, 50 mM sodium phosphate (NaP) containing 25 mM tetrabutylammonium (TBA) or octyl sulfate (C8) ion in 20% acetonitrile solution; flow rate, 1 ml min^{-1}; temperature, 40 °C; detection, UV 254 nm.*

Effect of the Ionic Strength of the Buffer Solution

In ion-exchange liquid chromatography, an increase in the buffer concentration or the addition of a salt is effective in improving the resolution. Increasing the inorganic ion concentration reduces the ion-pair formation in normal-phase ion-pair partition liquid chromatography. The retention factor values then become smaller. In contrast, the buffer concentration does not affect the retention of the paired-ion in reversed-phase ion-pair liquid chromatography. However, the solubility of inorganic salts in organic solvent-rich eluents is poor, making the maintenance of the instrument difficult. It is highly recommended that a low-concentration buffer solution should be used if the pH is not changed during the separation.

Selectivity of Organic Modifier and the Concentration Effect

The selectivity and the concentration effect of the organic modifier in the eluent are the same as those explained above for reversed-phase liquid chromatography. Increasing the modifier concentration reduces the retention factor, as does decreasing the counter-ion concentration. The degree of change in the retention factor is due to changes in the solubility of the solute in the organic solvent.

Column Temperature Effects

Increasing the column temperature reduces the retention factor. The ion-pair formation is based on a chemical equilibrium; therefore, temperature control is important to obtain reproducible results.

Detection

The molecular absorption intensity of polar compounds is usually small, but highly sensitive detection can be obtained after pre- or post-column derivatizations. The use of ultraviolet absorption or fluorescence-active counter-ions makes it possible to achieve highly sensitive detection of polar compounds and enhance the capability of ion-pair liquid chromatography. For example, *N,N*-dimethylprotriptyline has been used as a counter-ion for carboxylic acids[12] and picric acid for quaternary amines[13] in normal-phase ion-pair partition liquid chromatography. Phenethylammonium, cetylpyridinium, 1-phenethyl-2-pycolinium, and naphthalene-2-sulfonic acid have been used for sulfonic acid and alkyl amines detection.[14,15] Ion-pair post-column extraction was applied on-line for fluorescence detection.[16]

Ion-pair liquid chromatography can be applied to compounds separated by ion-exchange liquid chromatography, and mixtures of ionic and non-ionic compounds are easily separated. The latter separation is difficult by ion-exchange liquid chromatography. Anions can be separated by reversed-phase ion-pair liquid chromatography (Figure 4.18).

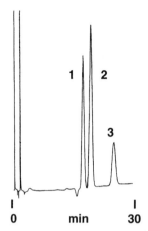

Figure 4.18 *Analysis of anions in water using ion-pair liquid chromatography. Column, octadecyl-bonded silica gel, 15 cm × 4.6 mm i.d.; eluent, 2 mM tetrabutyl-ammonium hydroxide (pH 5.3) in 3% acetonitrile–water; flow rate, 1 ml min^{-1}; detection, UV 200 nm. Peaks: 1, Br^-; 2, NO_3^-; and 3, I^-.*

Trouble-shooting in Ion-pair Liquid Chromatography

Ion-pair liquid chromatography can be applied to the separation of a wide variety of compounds. However, the flexibility of the selection of eluent components can confuse the operation. Trouble-shooting solutions are summarized in Table 4.6.

3 Normal-phase Liquid Chromatography

Normal-phase liquid chromatography was formerly called adsorption liquid chromatography. Pure or mixed organic solvents are used as the eluent and stationary phase adsorbent is more polar than the eluent. Hydrogen bonding is one of the important molecular interactions between sample molecules and the adsorbent. The possible molecular interactions are summarized in Table 1.1. When no molecular interaction is recognized, such chromatography is called size-exclusion liquid chromatography.

The basic molecular interaction in normal-phase liquid chromatography is electrostatic forces. The sample molecules are retained strongly by hydrogen bonding when the sample molecules themselves or the adsorbent act as both a hydrogen-bond acceptor and donor. The specificity of normal-phase liquid chromatography is the result of the direct formation of a strong molecular interaction between sample molecules and the adsorbent. The position of substitution of sample molecules directly affects the separation of isomers. The substituent effect is weak in reversed-phase liquid chromatography. Therefore, normal-phase liquid chromatography is suitable for the separation of isomers, such as *cis/trans, ortho/meta/para*, and steric isomers. The steric effect is especially important for chiral separations on suitable chiral columns. A

Table 4.6 *Trouble-shooting for ion-pair liquid chromatography*

Trouble	Reason	Solution
No peak found	Low conc. sample soln.	Check sensitivity of detector & adjust conc. sample soln.
	Unsuitable detector	Usually UV absorption of eluent is high because of counter-ion and buffer. Select suitable wavelength or change detector
	Disappeared in ghost peak (solvent, counter-ion)	Increase conc. sample soln. & compare with previous chromatogram. Or measure diode array chromatogram
	Adsorption in column	Change column Increase polarity of column Increase conc. organic modifier Dilute conc. counter-ion
	Leak from injector	Clean and fix injector
	Poor flow rate due to surfactant	Clean and fix pump head
	Poor detection due to surfactant	Clean and fix detector flow cell
No retention of analyte	Unsuitable column	Change column
	Unsuitable counter-ion	Select further hydrophobic counter-ion Increase conc. counter-ion
	Unsuitable conc. organic modifier	Decrease conc. organic modifier or don't use organic modifier
	Too high conc. buffer	Decrease conc. buffer, especially inorganic buffer
	Unsuitable pH	Change pH, take care with pK_a of analytes
Non-reproducible retention time	Unstable flow rate	Clean pump and flow line
	Unstable temperature	Use column heater or oven
	Non-equilibrated column	Equilibrate column, especially for initial injections
	Too diluted conc. buffer	Change buffer

Problem	Cause	Remedy
Peak tailing	Too strong adsorption on column	Change column
	Too dilute counter-ion	Increase conc. counter-ion
	Unsuitable eluent pH	Change pH suitable for ionization of analytes
	Non-equilibrated column	Equilibrate column, usually for 30–120 min
	Due to gradient	Use moderate gradient
	Overloaded sample	Reduce amount injected
Peak broadening	Low conc. organic	Increase conc. organic modifier
	Damaged column	Change column
	Unsuitable counter-ion	Use more hydrophobic counter-ion
	Unsuitable pH	Change pH of eluent
	High conc. basic counter-ion	Change column, use guard column
	pH > 7.5	Change column to suitable one, change eluent
Noise-like peak	Unstable flow rate (air, recrystallized salt, and counter-ion in pump head)	Clean pump head
	Recrystallization of salt counter-ion	Clean injector
High pressure drop		Change eluent, buffer components
		Wash column with water then methanol
	Due to injector or pump	Wash and fix injector and pump
	Too high conc. buffer	Dilute conc. buffer and counter-ion
Low sensitivity	Unsuitable wavelength	Select appropriate wavelength
	Dirty flow cell	Wash flow cell with water and methanol
	Unsuitable counter-ion	Change counter-ion to one having a chromophore or fluorophore and change the condition
	Aged lamp or detector	Fix detector

variety of chiral separation phases have been synthesized; however, one phase cannot separate many different types of optical isomers, because the steric effect also depends on the molecular structure of the analytes.

In ligand-exchange liquid chromatography, the separation is performed by the replacement of analytes, which form a complex with stationary phase materials or a reagent trapped on the stationary phase surface, by components in the eluent, which act as similar ligands to the analytes. In charge-transfer complex liquid chromatography, analytes are retained by charge-transfer complex formation on the surface of stationary phase material and are replaced by other components in the eluent which form a charge-transfer complex with the stationary phase materials or with reagents trapped on the stationary phase surface.

Normal-phase liquid chromatography is thus a steric-selective separation method. The molecular properties of steric isomers are not easily obtained and the molecular properties of optical isomers estimated by computational chemical calculation are the same. Therefore, the development of prediction methods for retention times in normal-phase liquid chromatography is difficult compared with reversed-phase liquid chromatography, where the hydrophobicity of the molecule is the predominant determinant of retention differences. When the molecular structure is known, the separation conditions in normal-phase LC can be estimated from Table 1.1, and from the solvent selectivity. A small-scale thin-layer liquid chromatographic separation is often a good tool to find a suitable eluent. When a silica gel column is used, the formation of a monolayer of water on the surface of the silica gel is an important technique. A water-saturated very non-polar solvent should be used as the base solvent, such as water-saturated *n*-hexane or isooctane.

Example 1: Chromatography of phthalic acid alkyl esters. Phthalic acid alkyl esters have been chromatographed on different types of stationary phase materials in normal-phase liquid chromatography. The stationary phase materials used were dimethylsilylated silica gel, propyl cyano-bonded silica gel, polystyrene gel, pellicular-type silica gel, and pellicular-type anion exchangers. In each case the longer the alkyl chain length of the phthalate, the shorter was the retention time. This result suggested that the elution order depended on the alkyl chain length of the samples and was not affected by the stationary phase materials, and that molecular interaction and solubility, not the type of stationary phase material, were important for the separation.[67]

Example 2: Chromatography of nitroaniline isomers. The elution order of the nitroaniline isomers was *ortho*, *meta*, and *para* in normal-phase liquid chromatography using *n*-butanol–*n*-hexane mixtures as the eluent, when the stationary phase material was either silica gel, alumina, an ion-exchanger, polystyrene gel, or octadecyl-bonded silica gel. The results indicate that the separation of these compounds can be performed on a range of different types of stationary phase materials if the correct eluent is selected. The best separation will be achieved by the right combination of stationary phase material and eluent.[68]

Classification of Solvents

The most practical method for finding a good eluent for normal-phase liquid chromatography is a trial experiment using thin-layer chromatography (TLC). When a rough separation is achieved by TLC and the maximum R_f value is less than 0.5, the developing solvent is a good candidate to be a suitable eluent for column liquid chromatography. The modification of components depends on the surface activity, *i.e.* the water content, and the surface area of silica gel.

The selection of the solvent can be based on the solvent properties classified by polarity (Table 4.7), hydrogen bonding (Table 4.8), and solubility parameters (Table 4.9). Snyder has classified solvents according to their hydrogen bonding acceptor, hydrogen bonding donor, and dipole properties.[70] The solvents useful for liquid chromatography are shown in Table 4.9. Water and chloroform are classified in the same group (VIII); however, these solvents are not miscible. The method for identifying a good developing solvent for TLC is also applicable to liquid chromatography. First a solvent in which the analyte is very soluble, based on the concept of 'like dissolves like' developed by Freizer, must be selected, then a solvent in which the analyte is not soluble is selected and these solvents have to be miscible together. The eluent is thus a mixture of solvents for the analyte.

The selection of a solvent is based on the solvent strength ε^0 experimentally obtained on silica gel and alumina.[69] An eluent having $\varepsilon^0 = 0.40$ on alumina can be made from a mixture of methyl acetate–pentane (19:81) or acetone–pentane (28:72). An eluent having $\varepsilon^0 = 0.20$ on alumina can be made from a mixture of acetone–pentane (6:94), chloroform–pentane (15:85), or benzene–pentane (28:72), as shown in Figure 4.19. However, no linear relationship between

Table 4.7 *Classification of solvents by Hecker*[70]

Non-polar solvent	
Hexane	Tetrahydrofuran
Heptane	Dioxan
Isooctane	Acetone
Cyclohexane	Aniline
Cyclopentane	Phenol
Carbon disulfide	Ethanol
Carbon tetrachloride	Glycol monomethyl ether
Benzene	Acetic acid
Trichloroethylene	Acetonitrile
Dichloroethylene	Nitromethane
Chloroform	Formic acid
Tetrachloroethylene	Morpholine
Dichloromethane	Formamide
Diethyl ether	Water
Methyl ethyl ketone	Inorganic acid
Ethyl acetate	▼ Salt, Buffer
▼ Pyridine	Polar solvent

Table 4.8 *Classification of solvent based on hydrogen bonding*

Strong donor & acceptor	Water, glycol, glycerol, aminoalcohol, hydroxylamine, polyphenol, and amide
Weak donor & acceptor	Alcohol, organic acid, phenol, primary and secondary amines, oxime, nitro and nitrile compounds having hydrogen at α-position, hydrazine, HF, and HCN
Donor	Ether, ketone, aldehyde, ester, *tert*-amines, pyridine, and nitro and nitrile compounds not having hydrogen at α-position
Acceptor	Chloroform, dichloroethane, CH_3CHCl_2, CH_2ClCH_2Cl, and $CH_2ClCHClCH_2Cl$
No hydrogen bonding	Hydrocarbons, carbon disulfide, mercaptans, solvents not included in above groups

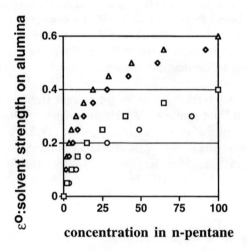

Figure 4.19 *Solvent strength of combination of n-pentane and more polar solvents in normal-phase liquid chromatography using alumina. Symbols:* △, *methyl acetate;* ◇, *acetone;* □, *chloroform; and* ○, *benzene.*

these mixed solvents is obtained; this is both a rough estimation method and a practical method.

The order of solvent strength (δ) (Table 4.9) and the adsorption parameter on alumina (ε^0) are not parallel. This indicates that solvent strength cannot be directly related to solvent selection in liquid chromatography. One way of using Table 4.9 in practice is to rearrange the solvents in terms of the values of ε^0. Increasing the ε^0 values strengthens the solubility. The retention time thus becomes shorter. This method is not perfect, and the elution power also depends on the structure of the analytes. Another method is the use of δ_d, δ_o, δ_a, and δ_h. These values indicate the interaction properties of the solvents. The solvent can be selected on the basis of these values and the properties of the analytes. The sum of these values and the steric effect may affect the chromatogram.

Table 4.9 *Solvents commonly used in liquid chromatography*

Solvent	δ	δ_d	δ_o	δ_a	δ_h	ε^0	η	VWV	M_r	d_m	RI	UV	BP (°C)
For normal-phase liquid chromatography													
Polyfluoroalkanes	6.0	6.0	0	1	0	−0.25	–	–	–	–	1.25	210	~50
Isooctane	7.0	7.0	0	0	0	0.01	0.50	101.93	114.230	0.015	1.3914	210	98
Diisopropyl ether	7.0	6.9	0.5	0.5	0	0.28	0.37	82.51	102.176	1.266	1.368	220	68
n-Pentane	7.1	7.1	0.00	0	0	0.00	0.23	67.02	72.150	0.020	1.3575	190	33
n-Hexane	7.3	7.3	0	0	0	0.01	0.313	77.30	86.177	0.016	1.3749	195	69
n-Heptane	7.4	7.3	0.0	0	0	0.01	0.41	89.70	100.203	0.006	1.3876	200	98
Diethyl ether	7.4	6.7	1.15	2	0	0.38	0.24	58.98	74.122	1.557	1.3524	218	34
Triethylamine	7.5	7.5	0	3.5	0	0.54	0.36	86.57	101.191	0.848	1.399	345	90
Cyclopentane	8.1	8.1	0.00	0	0	0.05	0.44	59.39	70.134	0.003	1.4064	200	49
Cyclohexane	8.2	8.2	0	0	0	0.04	1.00	70.89	84.161	0.000	1.4262	200	80
Propylchloride	8.3	7.3	3	0	0	0.30	0.35	51.18	78.541	1.725	1.4601	263	76
Carbon tetrachloride	8.6	8.6	0	0.5	0	0.18	0.97	53.86	153.823	0.001	1.4601	263	76
Ethyl acetate	8.6	7.0	1.88	2	0.5	0.58	0.45	57.15	88.106	1.769	1.3724	256	76
Propylamine	8.7	7.3	4	6.5	0	–	0.35	50.14	59.111	1.444	1.385	–	48
Chloroform	9.1	8.1	1.15	0.5	0	0.40	0.57	44.50	119.378	1.156	1.4458	245	60
Methyl acetate	9.2	6.8	4.5	2	0	0.60	0.37	45.29	74.079	1.725	1.362	260	56
Benzene	9.2	9.2	0	0.5	0	0.32	0.65	54.47	78.113	1.305	1.5011	278	80
Acetone	9.4	6.8	2.69	2.5	0	0.56	0.36	40.60	58.080	2.911	1.3587	330	56
Dichloromethane	9.6	6.4	1.14	0.5	0	0.42	0.44	36.14	84.933	1.503	1.4241	233	40
Propanol	10.2	7.2	3.09	4	4	0.82	2.3	46.13	60.096	1.534	1.3856	210	98
Pyridine	10.4	9.0	2.37	5	0	0.71	0.95	50.39	79.101	1.973	1.5102	330	115
Ethanol	11.2	6.8	4.0	5	5	0.88	1.20	34.24	46.069	1.551	1.361	210	78.5
Dimethylformamide	11.5	7.9	3.86	–	–	–	0.92	49.42	73.094	3.550	1.4305	268	153
Acetic acid	12.4	7.0	–	–	–	1.0	1.26	32.87	60.052	1.885	1.372	230	118

(continued)

Table 4.9 *Continued*

For reversed-phase liquid chromatography

Solvent	δ	δ_d	δ_o	δ_a	δ_h	ε^0	η	VWV	M_r	d_m	RI	UV	BP (°C)
Triethylamine	7.5	7.5	0	3.5	0	0.54	0.36	86.57	101.191	0.848	1.399	345	90
Propylamine	8.7	7.3	4	6.5	0.5	–	0.35	50.14	59.111	1.444	1.385	–	48
Tetrahydrofuran	9.1	7.5	1.75	3	0	0.45	0.55	40.87	68.075	0.492	1.4072	212	190
Acetone	9.4	6.8	2.69	2.5	0	0.56	0.36	40.60	58.080	2.911	1.3587	330	56
Dioxane	9.8	7.8	0.45	3	0	0.56	1.37	56.16	88.106	0.010	1.4224	215	101
Propanol	10.2	7.2	3.09	4	4	0.82	2.3	46.13	60.096	1.534	1.3856	210	98
Isopropanol	–	–	–	5	5	0.82	1.9	45.66	60.096	1.618	1.384	205	82
Ethanol	11.2	6.8	4.0	5	5	0.88	1.20	34.24	46.069	1.551	1.361	210	78.5
Dimethyl formamide	11.5	7.9	–	–	–	1.0	1.26	49.42	73.094	3.550	1.372	230	118
Acetonitrile	11.8	6.5	3.44	2.5	0	0.65	0.38	27.68	41.052	2.894	1.3441	190	82
Acetic acid	12.4	7.0	–	–	–	1.0	1.26	32.87	60.052	1.885	1.372	230	118
Trifluoroacetic acid	–	–	–	–	–	–	–	37.75	114.024	1.879	1.480	260	72.4
Dimethyl sulfoxide	12.8	8.4	3.9	5	0	0.6	2.24	48.43	78.129	3.948	1.483	260	190
Methanol	12.9	6.2	2.87	7.5	7.5	0.95	0.55	22.40	32.042	1.621	1.3284	205	65
Ethanolamine	13.5	8.3	L	L	L	–	–	41.57	61.083	1.166	1.4540	–	170
Formamide	17.9	8.3	L	L	L	–	–	24.42	45.041	3.699	1.4470	275	210
Water	21.0	6.3	L	L	L	1.00	1.00	9.38	18.015	1.860	1.333	190	100

δ: Total solubility parameter indicates solvent strength and polarity. Larger δ means polar, smaller means non-polar (hydrophobic), δ_d: the ability of a solvent to participate in dispersive interactions, which indicates degree of solubility of aromatic compounds having halogen and sulfur substituents. Larger values mean strong dispersive interaction. δ_o: orientation interaction; indicates degree of solubility of dipole compounds. δ_a: proton donor property; indicates degree of solubility of alcohol, phenol, and carboxylic acid. δ_h: proton acceptor property, indicates selective degree of solubility of basic compounds. ε^0: solvent strength in normal-phase liquid chromatography using alumina. η: viscosity (cp at 20°C), VWV*: van der Waals volume Å^3, M_r: relative molecular mass, d_m: dipole moment in debye, RI: refractive index, UV: UV cut-off (nm), BP: boiling point (°C), *calculated using CAChe™ program from Sony-Tektronix. This table is based on references by L.R. Snyder (ed.). 'Principles of adsorption chromatography', Marcel Dekker, New York, 1968, p. 174, and J.J. Kirkland (ed.), 'Modern practice of liquid chromatography', Wiley-Interscience, New York, 1971, p. 136.

For example, if the separation of a steroid mixture is difficult in one chromatographic system, the selection of a different system is needed to separate the mixtures. The substituents with a hydrogen-bonding effect due to a hydroxy group, a hydrogen-bonding donor effect of a carbonyl group and a hydrophobic effect due to an ester suggest the elution order and solvent selection. That is, the solvent selection is based on δ_d, δ_o, δ_a, and δ_h of analytes. Subsequently, another solvent that is miscible with the first solvent is used for dilution. The final eluent is a mixture of these two solvents.[71-73]

Preparation of the Eluent

'**Like dissolves like**' is the basic concept for the selection of solvents in the eluent for liquid chromatography. Controlling the solubility of analytes is the key to success. If the selected solvent or mixture of solvents does not interfere with detection, it is a good eluent. The selection of a suitable solvent for low-wavelength absorption detection and post-column derivatization detection is important to obtain highly sensitive detection. The selection of a volatile solvent is the key for preparative-scale liquid chromatography and for mass spectrometric detection.

The selection of the solvent is based on the retention mechanism. The retention of analytes on stationary phase material is based on the physicochemical interactions. The molecular interactions in thin-layer chromatography have been extensively discussed, and are related to the solubility of solutes in the solvent. The solubility is explained as the sum of the London dispersion (van der Waals force for non-polar molecules), repulsion, Coulombic forces (compounds form a complex by ion–ion interaction, *e.g.* ionic crystals dissolve in solvents with a strong conductivity), dipole–dipole interactions, inductive effects, charge-transfer interactions, covalent bonding, hydrogen bonding, and ion–dipole interactions. The steric effect should be included in the above interactions in liquid chromatographic separation.

One example is the separation of tricarboxylic acid cycle organic acids. These organic acids were originally separated on sulfonic acid-treated silica gel as the stationary phase material, and a chloroform and isopentyl alcohol mixture was used as the eluent. However, this eluent is not suitable for UV detection. A mixture of *n*-hexane, THF, and *tert*-butanol was therefore selected for the same separation.[74]

One component of the eluent should have properties similar to those of the analytes, and this solvent is diluted by another solvent to control the retention time. The basic idea can be understood from the chromatographic behaviour of phthalic acid esters and polycyclic aromatic hydrocarbons (PAH). This approach can be applied to the separation procedure for a variety of stationary phase materials, including silica gel, polystyrene gel, and ion-exchangers.

The elution order of phthalic esters is related to the carbon chain length. The longer the chain length, the shorter the retention time in normal-phase liquid chromatography, and the elution order is reversed in reversed-phase liquid chromatography. The elution order of PAHs is based on the number of double bonds, an increased number of double bonds leads to longer retention time in both normal- and reversed-phase liquid chromatographies. These elution orders are always observed on a variety of stationary phase materials, including ion-exchangers. The retention time can be modified by changing the eluent components or the ratio of the solvent mixture. Increasing the solubility in the eluent shortens the retention time in both normal and reversed-phase liquid chromatographies. Increasing the alkyl chain length of analytes makes them more similar to alkanes. The analyte becomes more hydrophobic (lipophilic) and hence soluble in *n*-hexane, but not in water. The solubility of both short and long alkyl chain compounds is good in ethanol, which is miscible with *n*-hexane and water. The addition of ethanol to the eluent makes the retention time shorter in both normal and reversed-phase liquid chromatographies. This phenomenon can be observed in the chromatographic behaviour of aliphatic acids. However, the solubility of PAHs is poor even in ethanol; therefore, the elution order is always the same in both normal- and reversed-phase liquid chromatographies.

In normal-phase liquid chromatography, the elution order of benzene derivatives on silica gel is alkylbenzenes with a long alkyl chain < alkylbenzenes with a shorter alkyl chain < benzene < fluorobenzene < chlorobenzene < cyanobenzene < nitrobenzene ≪ aniline < phenol < benzoic acid. The elution order of aniline derivatives is dinitroaniline < benzene < anisidine < chloro-aniline < nitroaniline < phenylenediamine < aminophenol. This elution order is also observed for anion-exchangers. Alkanes are almost unretained. Alkenes are retained according to the number of double bonds. The elution order for *cis* and *trans* isomers is *cis,cis* < *cis,trans* < *trans,trans*. This also suggests that the eluent component should be selected according to the δ_o of dipole moment, δ_a of hydrogen bonding acceptor, and δ_h of hydrogen bonding donor, but not the polarity δ.

In other interactions the surface silanol groups of silica gel can form hydrogen bonds, and an alumina surface can form hydrogen bonds and a charge-transfer complex. However, such molecular interactions are caused by positive and negative charge sites, as in Lewis acid–base interactions. The chloro-ion of chloride-form anion-exchangers may form hydrogen bonds with aniline and phenol. Silver ions of silver-form cation-exchangers form charge-transfer complexes with the π-electrons of alkenes and aromatic compounds.

4 Size-exclusion Liquid Chromatography (SEC)

This separation method is based on the molecular size of analytes. Analytes pass through porous stationary phase materials having different pore sizes, and molecular interactions between analytes and the stationary phase surface must be eliminated. A very strong solvent is therefore required in this system. This system is also called gel filtration liquid chromatography, gel-permeation liquid chromatography, or molecular sieve chromatography. This system is used to measure the average molecular mass and for the purification of large- from small-size molecules.

Aqueous Phase Size-exclusion

The elution time is related to the logarithmic relative molecular mass (log M_r) within a certain range of M_r for a particular column, as seen in Figure 4.20. The relationship for proteins is not as simple as that for dextran and polyethylene glycol. This is because the molecular shape of proteins differ even though 0.2 M sodium phosphate buffer, pH 6.8, was used as the solvent. The molecular shape effect can be understood from the different calibration curves in Figure 4.20. The elution time of polyethylene glycol, whose structure is like a straight chain, was longer than that of dextran. Dextran and polyethylene glycol can be eluted with only water as the solvent. The hydrolysates of dextran have been separated by SEC in pure water.[75]

The non-uniform behaviour of proteins makes the measurement of their molecular mass difficult; however, their rough estimation and purification are very important for biochemical research. Generally, phosphate buffer and tris–HCl buffer (pH 7) are used for SEC for biological polymers, and some additives are required for further separation. Only 0.2 M sodium phosphate buffer was

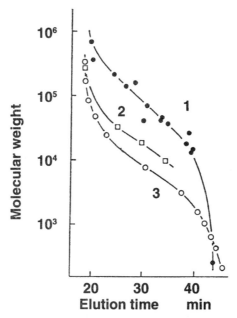

Figure 4.20 *Calibration curves for size-exclusion liquid chromatography. Column, TSK GEL G3000SW, 120 cm × 7.5 mm i.d.; eluent, 0.2 M sodium phosphate buffer pH 6.8; flow rate, 1.0 ml min^{-1}. Standards: 1, protein; 2, dextran; and 3, polyethylene glycol.*
(Reproduced by permission from Tosoh data)

used for the analysis of gelatin. However, 0.1 M sodium phosphate buffer (pH 7.0) containing 0.1 M sodium chloride was required for the analysis of ribonucleic acid from *E. coli*,[7] and 0.1 M sodium phosphate buffer (pH 7.0) containing 0.3 M sodium chloride was required for that of serum from bone marrow.[8] Tris–HCl buffer (pH 7.5) containing 0.1% sodium dodecylsulfate was used for the analysis of cellular membrane protein from rat microsomes,[7] and 50 mM Tris–HCl buffer (pH 7.2) containing 2 mM EDTA and 300 mM NaCl for the analysis of plasmid DNA fragments as the elution solvent.[8]

SEC is a useful tool for monitoring enzyme reactions, as seen in Figure 4.21, where the speed of decomposition of β-lactoglobulin by α-chymotrypsin is shown. SEC is widely used for purification of proteins, but the separation is due only to the difference in molecular mass. Therefore, ion-exchange liquid chromatography is combined with SEC to improve the selectivity.

Organic Phase Size-exclusion

The average molecular mass of synthesized polymers can be measured by SEC. This system has been called gel-permeation liquid chromatography, using porous polystyrene gel as the stationary phase material. Polymers are usually dissolved in THF or chloroform for separation by SEC. The molecular density

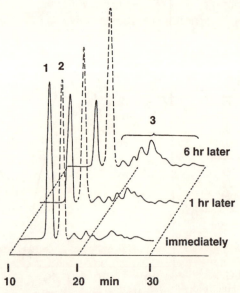

Figure 4.21 *Monitoring of an enzyme reaction using size-exclusion liquid chromatography. Column, TSK GEL G3000SW, 60 cm × 7.5 mm i.d.; eluent, 0.07 M potassium phosphate buffer containing 0.1 M potassium chloride; flow rate, 1 ml min⁻¹; detection, UV 280 nm. Peaks: 1, β-lactoglobulin; 2, α-chymotrypsin; and 3, decomposed products.*
(Reproduced by permission from Tosoh data)

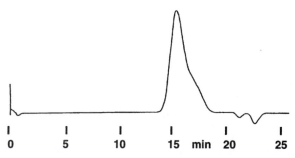

Figure 4.22 *High temperature size-exclusion liquid chromatography of an engineering plastic, poly(phenyl sulfate). Column, SSC GPS-3506, 50 cm × 8 mm i.d.; eluent, 1-chloronaphthalene; flow rate, 1.0 ml min⁻¹; column temperature, 210 °C; detector, refractive index detector.*
(Reproduced by permission from Senshu Kagaku data)

of analytes is measured by comparison of chromatograms obtained from light scattering and refractive index detectors.

Newly synthesized organic polymers having high temperature resistance and physical strength are not soluble in THF or chloroform. High-temperature SEC is therefore being developed, and the pressure-drop measurement detector and flame ionization detector are also used for monitoring the high temperature effluent. A chromatogram of an engineering plastic, poly(phenyl sulfate), is shown in Figure 4.22. The column temperature was 210 °C and the eluent was 1-chloronaphthalene.[77]

5 References

1. L.R. Snyder, *J. Chromatogr. Sci.*, 1978, **16**, 223.
2. D.J. Pietrzyk and C.-H. Chu, *Anal. Chem.*, 1977, **49**, 860.
3. T. Hanai, K.C. Tran, and J. Hubert, *J. Chromatogr.*, 1982, **239**, 385.
4. T. Hanai and J. Hubert, *HRC & CC*, 1984, **7**, 524.
5. Y. Arai, M. Hirukawa, and T. Hanai, *Nihon Kagaku Kaishi*, 1986, 969.
6. From the Mitsubishi Kasei catalogue.
7. From the Tosoh catalogue.
8. From the Asahipak catalogue (Showa Denkko).
9. G. Schill, R. Modin, and B.-A. Person, *Acta Pharm. Seuc.*, 1965, **2**, 119.
10. B.L. Karger, S.C. Su, S. Maichese, and B.-A. Person, *J. Chromatogr. Sci.*, 1974, **12**, 678.
11. M. Hirukawa, M. Maeda, A. Tsuji, and T. Hanai, *J. Chromatogr.*, 1990, **507**, 95.
12. S. Eksborg, P.-O. Lagerstrom, R. Modin, and G. Schill, *J. Chromatogr.*, 1993, **83**, 99.
13. E. Eksborg and G. Schill, *Anal. Chem.*, 1973, **45**, 2029.
14. M. Denkert, L. Hackzell, G. Schill, and E. Sjogren, *J. Chromatogr.*, 1981, **218**, 31.
15. L.E. Vera-Avla, M. Caude, and R. Rosset, *Analusis*, 1982, **10**, 43.
16. K. Sugden and G.B. Loscombe, *J. Chromatogr.*, 1978, **149**, 377.
17. B.A. Bildingmeyer and F.V. Warren, *Anal. Chem.*, 1984, **54**, 2351.
18. S.J. Valenty and P.E. Behnken, *Anal. Chem.*, 1978, **50**, 834.
19. B. Franssod, K.-G. Wahlund, I.M. Johansson, and G. Schill, *J. Chromatogr.*, 1976, **125**, 327.

20. J.F. Lawrence, F.E. Lancaster, and H.B. Conacher, *J. Chromatogr.*, 1981, **210**, 168.
21. M.G.M. de Ryter, R. Cronnelly, and N. Castagnoli, Jr., *J. Chromatogr.*, 1980, **183**, 193.
22. W. Voelter, K. Zech, P. Arnold, and G. Ludwig, *J. Chromatogr.*, 1980, **199**, 345.
23. D.N. Buchanan, F.R. Fucek, and E.F. Domino, *J. Chromatogr.*, 1980, **181**, 329.
24. R.G. Peterson, B.H. Rumack, J.B. Sullivan, and A. Makowski, *J. Chromatogr.*, 1980, **188**, 420.
25. M. Ghias-Ud-Din, E.B. Olson, Jr., and J. Rankin, *J. Chromatogr.*, 1980, **192**, 463.
26. K.T. Muir, J.H.G. Jonkman, D.-S. Tang, M. Kunitani, and S. Riegelman, *J. Chromatogr.*, 1980, **221**, 85.
27. M. Schoneshofer and A. Fenner, *J. Chromatogr.*, 1981, **224**, 472.
28. C.M. Riley, L.A. Sternson, and A.J. Repta, *J. Chromatogr.*, 1983, **276**, 93.
29. M. Kneczke, *J. Chromatogr.*, 1980, **198**, 529.
30. M. Alvinerie, P. Galtier, and G. Escoula, *J. Chromatogr.*, 1981, **223**, 445.
31. N. Kurosawa, S. Morishima, E. Owada, and K. Ito, *J. Chromatogr.*, 1984, **305**, 485.
32. D. Beales, R. Finch, and A.E. McLean, *J. Chromatogr.*, 1981 **226**, 498.
33. P.B. Baker and T.A. Gough, *J. Chromatogr. Sci.*, 1981, **19**, 483.
34. J. Wagner, M. Palpeyman, and M. Zraika, *J. Chromatogr.*, 1979, **164**, 41.
35. R.V. Smith, J.C. Glade, and D.W. Humphrey, *J. Chromatogr.*, 1979, **172**, 520.
36. K. Makino, H. Ozaki, T. Matsumoto, H. Imaishi, and T. Takeuchi, *J. Chromatogr.*, 1987, **400**, 271.
37. Y. Kobayashi, H. Kubo, and T. Kinoshita, *J. Chromatogr.*, 1987, **400**, 113.
38. G. Inchauspe, P. Delrieu, P. Dupin, M. Laurent, and D. Samain, *J. Chromatogr.*, 1987, **404**, 53.
39. I. de Miguel, E. Puech-Costes, and D. Samain, *J. Chromatogr.*, 1987, **407**, 109.
40. J.J. Lauff, *J. Chromatogr.*, 1987, **417**, 99.
41. K. Lindgren, *J. Chromatogr.*, 1987, **413**, 351.
42. G. Lam, R.M. Williams, and C.C. Whitney, *J. Chromatogr.*, 1987, **413**, 309.
43. D.L. Theis, *J. Chromatogr.*, 1987, **402**, 335.
44. A. Sokolowski and K.-G. Wahlund, *J. Chromatogr.*, 1980, **189**, 299.
45. Y. Yoshitoku, M. Moriyasu, and Y. Hashimoto, *Chem. Lett.*, 1982, 193–194.
46. P.G. Rigas and D.J. Pietrzyk, *Anal. Chem.*, 1987, **59**, 1388.
47. B.A. Bildingmeyer, C.T. Santasania, and F.V. Warren, Jr., *Anal. Chem.*, 1987, **59**, 1843.
48. L. Gagliardi, G. Cavazzutti, A. Amato, A. Basili, and D. Tonelli, *J. Chromatogr.*, 1987, **394**, 345.
49. M.L. Abaugh, I.S. Krull, and W.R. Lacourse, *J. Chromatogr.*, 1987, **387**, 301.
50. J.H. McB. Miller, C. Pascal, and M. Tissieres, *J. Chromatogr.*, 1987, **392**, 361.
51. J.A. De Shutter, W. Van Den Bossche, and P. De Moerloose, *J. Chromatogr.*, 1987, **391**, 303.
52. H. Lentzen and R. Simon, *J. Chromatogr.*, 1987, **389**, 444.
53. E. Naline, B. Flouvat, C. Advenier, and M. Pays, *J. Chromatogr.*, 1987, **419**, 177.
54. J.H. Kennedy and B.A. Olsen, *J. Chromatogr.*, 1987, **389**, 369.
55. M. Dreux, M. Lafosse, P. Agbo-Hazoume, B. Chaabane-Doumandji, M. Gibert, and Y. Levi, *J. Chromatogr.*, 1986, **354**, 116.
56. S.P. Sood, L.E. Sartori, D.P. Wittmer, and W.G. Haney, *Anal. Chem.*, 1976, **48**, 796.
57. K.-G. Wahlund and A. Sokolowski, *J. Chromatogr.*, 1978, **151**, 299.
58. N.D. Brown and H.K. Sleeman, *J. Chromatogr.*, 1977, **138**, 449.
59. D.P. Wittmer, N.O. Nuessle, and W.G. Haney, Jr., *Anal. Chem.*, 1975, **47**, 1422.
60. E. Fitzgerald, *Anal. Chem.*, 1976, **48**, 1734.
61. J.H. Knox and J. Jurand, *J. Chromatogr.*, 1975, **110**, 103.
62. J.C. Kraak, K.M. Jonker, and J.F.K. Huber, *J. Chromatogr.*, 1977, **142**, 671.
63. K.-G. Wahlund and U. Lund, *J. Chromatogr.*, 1976, **122**, 269.
64. J.H. Knox and G.R. Laird, *J. Chromatogr.*, 1976, **122**, 17.
65. K.-G. Wahlund, *J. Chromatogr.*, 1973, **115**, 411.

66. J. Mitchell and C.J. Coscia, *J. Chromatogr.*, 1978, **145**, 295.
67. W. Funasaka, T. Hanai, and K. Fujimura, *J. Chromatogr. Sci.*, 1974, **12**, 517.
68. T. Hanai and K. Fujimura, *J. Chromatogr. Sci.*, 1976, **14**, 140.
69. R. Neher, in 'Thin-layer chromatography', ed. G.B. Marini-Bettola, Elsevier, Amsterdam, 1964, p. 75.
70. L.R. Snyder, *J. Chromatogr.*, 1974, **92**, 223.
71. S. Hara, *J. Chromatogr.*, 1977, **137**, 199.
72. S. Hara, Y. Fujii, M. Hirasawa, and S. Miyamoto, *J. Chromatogr.*, 1978, **149**, 143.
73. S. Hara and A. Ohsawa, *J. Chromatogr.*, 1980, **200**, 85.
74. H. Hyakutake and T. Hanai, *J. Chromatogr.*, 1975, **108**, 385.
75. From Shodex data.
76. From Senshukagaku data.

Bibliography of additional references on theoretical approaches in ion-pair liquid chromatography

A. Bartha, H.A.H. Billiet, L.D. Galan, and G. Vigh, 'Studies in reversed-phase ion-pair chromatography III. The effect of counter-ion concentration', *J. Chromatogr.*, 1984, **291**, 91.

A. Bartha, G. Vigh, H. Billiet, and L. de Galan, 'Effect of the type of ion-pairing reagent in reversed-phase ion-pair chromatography', *Chromatographia*, 1985, **10**, 587.

B.A. Bidlingmeyer, S.N. Deming, W.P. Price, Jr., B. Sachok, and M. Petrusek, 'Retention mechanism for reversed-phase ion-pair liquid chromatography', *J. Chromatogr.*, 1979, **186**, 419.

J.B. Green, 'Liquid chromatography on silica using mobile phases containing tetraalkyl-ammonium hydroxides', *J. Chromatogr.*, 1986, **358**, 53.

M.T.W. Hearn (ed.), 'Ion-pair Chromatography', Marcel Dekker, New York, 1985.

S.O. Jansson, 'Effect of counter-ions in ion-pair liquid chromatography of hydrophobic amines on non-polar bonded phases', *J. Liq. Chromatogr.*, 1982, **5**, 677.

J.C. Kraak, H.H. van Rooij, and J.L.G. Thus, 'Reversed-phase ion-pair systems for the prediction of *n*-octanol/water partition coefficients of basic compounds by high-performance liquid chromatography', *J. Chromatogr.*, 1986, **352**, 455.

W.R. Melander and Cs. Horváth, 'Mechanistic study of ion-pair reversed-phase chromatography', *J. Chromatogr.*, 1980, **201**, 211.

M. Puttemans, L. Dryon, and D.L. Massart, 'Extraction of organic acids by ion-pair formation with tri-*N*-octylamine', *Anal. Chim. Acta*, 1984, **161**, 221.

W. Santi, J.M. Huen, and R.W. Frei, 'High-speed ion-pair partition chromatography in pharmaceutical analysis', *J. Chromatogr.*, 1975, **115**, 423.

G. Schill, 'Selective extraction of organic compounds as ion-pairs and adducts', *Talanta*, 1975, **22**, 1017.

M. Tang and S.N. Deming, 'Interfacial tension effects of nonionic surfactants in reversed-phase liquid chromatography', *Anal. Chem.*, 1983, **55**, 425.

O.A.G.J. van der Houwen, R.H.A. Sorel, A. Hulshoff, J. Teeuwsen, and A.W.M. Indemans, 'Ion-exchange phenomena and concomitant pH shifts on the equilibration of reversed-phase stationary phase materials with ion-pairing reagents', *J. Chromatogr.*, 1981, **209**, 393.

CHAPTER 5

Separation Based on an Improved Column Efficiency

High-speed separation can be achieved by reducing the column length, column internal diameter, the size of the stationary phase particles, and by increasing the eluent flow velocity, as explained in Chapter 1. In addition, the development and optimization of pumps, detectors, injectors, and the recording systems contribute to improved separations. The influence of diffusion theory on the experimental separation should also be considered. If the unresolved chromatogram shown in Figure 5.1A has been obtained for two compounds, two methods can be applied to achieve a complete separation. Figure 5.1B demonstrates that achieved by improving selectivity (relative retentions), and Figure 5.1C demonstrates a separation achieved by improving the column efficiency.

The resolution, R_s, is used to describe the separation conditions of two compounds. The R_s value is given by the ratio of the distance between

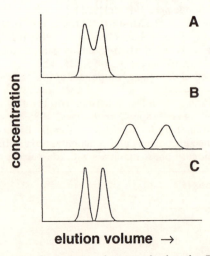

Figure 5.1 *Improvement of separation. A, unresolved peaks; B, increase in peak separation; C, improved peak efficiency.*

the maxima of two peaks, Δt_R, and the mean of the peak widths, $\bar{w}_b$, at their base:

$$R_s = \frac{t_{R2} - t_{R1}}{(w_{b2} + w_{b1})/2} = \frac{\Delta t_R}{\bar{w}_b} \tag{5.1}$$

where t_{R1} and t_{R2} are the retention times of the peaks, and w_{b1} and w_{b2} are the base widths of the two peaks.

Theoretically, when R_s is 1, the two peaks are 98% separated. An R_s value > 1 indicates a good separation and an R_s value < 1 indicates a poor separation. In Figure 5.2, the resolution between peaks 3 and 4 was 1.1. The resolution between peaks 4 and 5 was 1.5, and that between peaks 5 and 6 was 2.4. The resolutions do not exactly match with their respective peak shapes because of their different intensities. If two symmetric peaks have a separation that results in a resolution of $R_s = 1$, when the peaks have equal

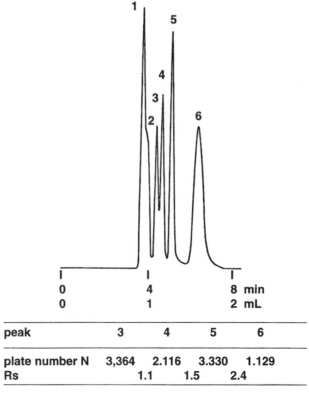

peak	3	4	5	6
plate number N	3,364	2.116	3.330	1.129
Rs		1.1	1.5	2.4

Figure 5.2 *Normal phase separation of phthalates. Column, Yanapak CN, 25 cm × 2 mm i.d.; eluent, n-hexane–n-butanol (200:1); flow rate, 0.25 ml min^{-1}; pressure, 3 MPa. Peaks: 1, lauryl phthalate; 2, heptyl phthalate; 3, butyl phthalate; 4, propyl phthalate; 5, ethyl phthalate; and 6, methyl phthalate.*

areas (Figure 5.3A), the resolution will deteriorate if there is a change in the ratio of the areas of the peaks, even if the retention times remain constant. An example is shown in Figure 5.3B; as the ratio increases the experimental resolution decreases. In designing a procedure it is therefore desirable to obtain equal peak areas for overlapping peaks. When the wavelength of the isosbestic point of the spectra of two compounds is used for photometric detection, the peak areas of a 1:1 mixture will be equal. Alternatively, the selection of the wavelength can be used to generate equal peak areas even if the mixture ratio is very different. When pure standard chemicals are used, their spectra should first be measured, and the most suitable wavelength selected to obtain good resolution and the correct peak area ratio. In practice, the selectivity depends on the structure of the analytes and the mixture ratio, and it is therefore often difficult to obtain an appropriate relationship between the R_s value and the mixture ratio. When one large peak is observed on a chromatogram, the purity of the peak should be studied by using a longer column or changing the detector wavelength. Conversely, poor resolution on a chromatogram can be improved by appropriate selection of the detector wavelength.

An improvement in the resolution can also be achieved by increasing the separation of the peak (Δt_R) or narrowing the peak width (increased N). An increase in Δt_R depends on the selection of stationary phase materials and eluent; a reduction of the peak width depends on improving the column

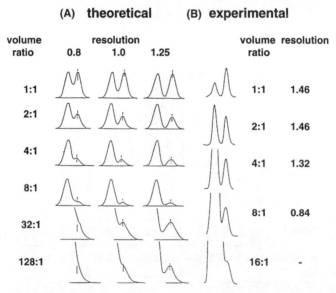

Figure 5.3 *Resolution related to peak area ratio.* (A) *theoretical and* (B) *experimental changes in resolution with volume ratio. Column,* 10 *μm octadecyl-bonded silica gel,* 15 *cm* × 4.6 *mm i.d.; eluent,* 90% *aqueous acetonitrile; flow rate,* 1 *ml* min^{-1}; *detection,* UV 254 *nm. Sample: different ratios of* 10 *μl benzene and toluene in* 10 *ml ethanol.*

efficiency. The enhancement of high-performance liquid chromatography depends mainly on improving the column efficiency.

1 Improving Separation by Changing the Selectivity

The selectivity of a separation is based on differences in the physicochemical interactions of the analytes with the stationary and mobile phases. This difference is measured as Δt_R on a chromatogram.

The retention volume of each component, V_R, is given by:

$$V_R = V_M + K \times V_S \tag{5.2}$$

where V_M is the volume of mobile phase inside the column, V_S is the volume of the stationary phase in reversed-phase chromatography, and K is the distribution constant of the analyte between the mobile and stationary phases. In normal-phase liquid chromatography, V_S is related to the surface area of the stationary phase material, and in ion-exchange liquid chromatography to the ion-exchange capacity. Differences in K values between analytes in one system reflect differences in the properties of the compounds. In any one column the distribution between the stationary and mobile phases is described by the mass distribution coefficient. To change values of ΔK between analytes, either the phases must be changed (changes in stationary and mobile phase selectivity, see earlier) or the temperature must be changed.

$$\text{Mass distribution constant} = \frac{\text{amount of solute in stationary phase}}{\text{amount of solute in mobile phase}}$$

$$= K \frac{V_S}{V_M}$$

where V_S and V_M are constant for all analytes in a particular column. This value is equal to the retention factor k. The stronger is the interaction between the sample molecule and the stationary phase the larger the retention of solute. Thus the larger the difference in K between analytes (ΔK), the better will be their separation. To improve the separation, the difference between the distribution constants (K values) of compounds 1 and 2 must be increased, or the volume of the stationary phase (V_S) must be increased.

$$\Delta t_R = V_{R2} - V_{R1} = \frac{V_M + K_2 V_S}{V_M + K_1 V_S} = V_S(K_2 - K_1)$$

To increase V_S, the chromatographer can increase the surface area of the stationary phase materials in normal-phase liquid chromatography, increase the stationary phase volume in reversed-phase or partition liquid chromatography, or increase the ion-exchange capacity in ion-exchange liquid chromatography. In general, if the internal diameter of a column is constant, the retention time difference (Δt_R) is related to the column length (L). Therefore, doubling the column length doubles the difference Δt_R. However, the resolution of the analyte (R_s) is only related to $\sqrt{L}$ due to the diffusion of the samples inside the column.

There is no molecular interaction in size-exclusion liquid chromatography, and therefore the resolution can only be improved by increasing the column length.

2   Improving Separation by Increasing the Column Efficiency

The peak broadening due to diffusion is related to the square root of L as follows:

$$w_b \propto \sqrt{HL} \tag{5.3}$$

where H is the column efficiency, *i.e.* the height equivalent to a theoretical plate (HETP). A smaller H value gives better resolution.

From Equation 5.3:

$$H \propto w_b^2 / L \tag{5.4}$$

If a peak diffuses as a theoretical Gaussian distribution, w_b is equal to 4σ. One peak width is 4σ standard deviation, from Equation 5.4:

$$H \propto \sigma^2 / L \tag{5.5}$$

The separation factor α of two compounds is given from Figure 3.4:

$$\alpha = \frac{t_{R2} - t_{R0}}{t_{R1} - t_{R0}} = \frac{V_{R2} - V_0}{V_{R1} - V_0} = \frac{K_2}{K_1} \tag{5.6}$$

The number of theoretical plates is given by:

$$N = 16\left(\frac{V_R}{w_b}\right)^2 = 16\left(\frac{t_R}{w_b}\right)^2 = 5.545\left(\frac{t_R}{w_h}\right)^2 = \left(\frac{t_R}{\sigma}\right)^2 \tag{5.7}$$

The effective number of theoretical plates is the number of usable theoretical plates and is calculated after elimination of the void volume:

$$N_{eff} = \left(\frac{t_R - t_0}{\sigma}\right)^2 = \left(\frac{k}{1+k}\right)^2 N \tag{5.8}$$

If the peak widths of peaks 1 and 2 are assumed to be equal, the following equation is derived from Equation 5.1:

$$R_s = \frac{2\Delta t_R}{w_{b1} + w_{b2}} = \frac{2(V_{R2} - V_{R1})}{2w_{b2}} = \frac{V_{R2}}{w_{b2}} \times \frac{V_{R2} - V_{R1}}{V_{R2}}$$

$$= \frac{V_{R2}}{w_{b2}} \times \frac{(V_{R2} - V_M) - (V_{R1} - V_M)}{V_{R2} - V_M} \times \frac{V_{R2} - V_M}{V_M} \times \frac{V_M}{V_{R2}}$$

(5.9)

From Equation 5.7,

$$\frac{V_{R2}}{w_{b2}} = 0.25\sqrt{N}$$

(5.10)

From Equation 5.2, $(V_{R2} - V_M) = K_2 V_S$, $(V_{R1} - V_M) = K_1 V_S$
Hence:

$$R_c = \frac{(V_{R2} - V_M) - (V_{R1} - V_M)}{V_{R2} - V_M} = \frac{K_2 V_S - K_1 V_S}{K_2 V_S} = \frac{K_2 - K_1}{K_2}$$

From Equation 5.6, $\alpha = K_2/K_1$, therefore

$$R_s = \frac{(K_2/K_1) - 1}{K_2/K_1} = \frac{(\alpha - 1)}{\alpha}$$

The mass distribution coefficient $k = K_c V_S/V_M = (V_R - V_M)/V_M$, and

$$\frac{V_M}{V_{R2}} = \frac{1}{V_{R2}/V_M} = \frac{1}{(V_{R2} - V_M)/V_M + 1} = \frac{1}{k + 1}$$

Therefore, from Equation 5.9:

$$R_s = \frac{\sqrt{N}}{4} \times \frac{\alpha - 1}{\alpha} \times k \times \frac{1}{1 + k} = \frac{1}{4}\left(\frac{\alpha - 1}{\alpha}\right)\left(\frac{k}{1 + k}\right)\sqrt{N}$$

(5.11)

Combination of Equations 5.8 and 5.11 gives:

$$R_s = \frac{1}{4}\left(\frac{\alpha - 1}{\alpha}\right)\sqrt{N}$$

(5.12)

The relationships between R_s, α, and N are given in Figure 5.4. When α is small, a large N is required for a good separation. However, the use of a longer column should be avoided for fast analysis even if a longer column provides excellent plate numbers. An improvement in the value of α can be estimated for a good separation.

A column having a smaller HETP value is a good column because diffusion inside the column is small, resulting in better separation. The HETP value is given by the Van Deemter equation, which describes the peak broadening of packed columns through which a non-compressible solvent is moving.

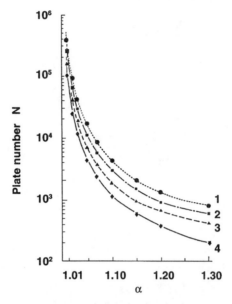

Figure 5.4 *Relationship between separation factor α, resolution R_s, and plate number N.*
Line: 1, R_s = 1.5; 2, R_s = 1.25; 3, R_s = 1.0; and 4, R_s = 0.8.

$H = H_F + H_L + H_S + H_M$. The H value can be expressed as the sum of (a) eddy diffusion (H_F), (b) longitudinal diffusion (H_L), and (c) resistance to mass transport in stationary and mobile phases (H_S and H_M, respectively).

Reducing the Eddy Diffusion Term

Eddy diffusion is a result of the presence of particles of stationary phase material in a column, and depends on the stationary phase conditions, shape of the column, and the structure of the stationary phase material. The influence of the stationary phase material can be divided into the particle size (d_p), the shape of the particles, and the porosity of the particles. The standard deviation for peak broadening due to the particles is described by: $\sigma_F{}^2 = 2\lambda d_p L$, and therefore from Equation 5.5:

$$H_F = 2\lambda d_p \tag{5.13}$$

λ depends on the irregularity of the particles (particle shape) and on the column material (steel gives more disturbance than glass); the relative effect will always increase as the column diameter decreases. The above-mentioned effects have led to an optimum internal diameter of 2–5 mm for LC columns. In a column with this internal diameter, which is uniformly packed with small spherical particles, the eddy diffusion will be limited. The λ value is about 1 for spherical particles.

Reducing the Effect of Longitudinal Diffusion

Molecules tend to diffuse randomly, in no particular direction, within any fluid, independently of the flow rate of the mobile phase. Their diffusion rate is determined by the type of molecule, the nature of the mobile phase, and the temperature, and is expressed quantitatively by their diffusion constants.

Molecular diffusion is related to time, as shown by the Einstein equation:

$$\sigma^2 = 2Dt \tag{5.14}$$

where D is the diffusion coefficient. The standard deviation due to longitudinal diffusion is σ_L, and the time during which diffusion can occur is $t_a = L/u$; therefore, $\sigma_L^2 = 2\gamma D_M t_a = (2\gamma D_M L)/u$, and from Equation 5.5, $H = \sigma^2/L$

$$H_L = (2\gamma D_M)/u \tag{5.15}$$

where γ is the factor of anti-diffusion, and is less than 1; D_M is the diffusion constant of sample molecules in the mobile phase; and u is the linear velocity of the mobile phase. The term H_L is important in gas chromatography because D_M is high. The D_M value in liquids is *ca.* 10^{-5}, and the flow rate is greater than the diffusion speed; therefore, H_L is not as important.

The H_L value is reduced by an increase in the viscosity of the solvent or by a decrease in the temperature. Longitudinal diffusion can thus be reduced by decreasing the diffusion coefficient and increasing the flow rate; however, these two actions are counter-effective in liquid chromatography because of the mass transport term.

Reducing the Effect of Resistance to Mass Transport in the Stationary and Mobile Phases

This dispersion effect results from the mass transfer of the analyte between the stationary and mobile phases, and is an important phenomenon in high-performance liquid chromatography. The injected analyte molecules are first present in the mobile phase. They then transfer back and forth from the stationary phase in order to interact. This process is repeated along the column, from the inlet to outlet. Such a phenomenon occurs in all kinds of chromatographies: normal-phase, reversed-phase, partition and ion-exchange liquid chromatography. (In size-exclusion liquid chromatography, such movement of sample molecules occurs as a diffusion process between the inside and outside of the stationary phase material.) Under ideal conditions, the sample molecules move along the column at the same speed in both the mobile and stationary phases (dotted lines in Figure 5.5). In practice, the movement of the mobile phase drags the analyte along faster than their movement through the stationary phase because equilibration at the interface is restricted by the diffusion rates within the phases. Therefore peak broadening occurs according to the total diffusion profile, as shown in Figure 5.5. The total diffusion effect

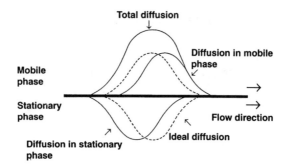

Figure 5.5   *Diffusion of analyte in mobile and stationary phases while moving along the column.*

can be divided into diffusion in the stationary phase (H_S) and that in the mobile phase (H_M). Within a certain time (t_a), the sample molecules in the mobile phase move a particular distance (ut_a) at a linear velocity (u). During this period, the movement of the sample molecules in the stationary phase is zero. If the number of sample molecules transferred between the mobile and stationary phases is defined as n_s, and the distance the centre of the band has moved is defined as d_R; n_s can be related to $2d_R/ut_a$. The time the analyte spends in the stationary phase is t_d. The ratio R of the moving speed of the band and the mobile phase is defined as $t_a/(t_a + t_d)$. The following equation is then obtained:

$$n_s = 2L/ut_a = [2(1 - R)L]/Rut_d$$

The distance (l_s) that the band centre has moved is related to R and ut_d:

$$l_s = Rut_d = (1 - R)ut_a$$

The standard deviation caused by molecular diffusion is:

$$\sigma^2 = l_s^2 n_s \qquad (5.16)$$

Therefore, the standard deviation due to molecular diffusion inside the stationary phase (σ_S) is:

$$\sigma_S^2 = 2qR(1 - R)uLt_d \qquad (5.17)$$

where q is a parameter that depends on the type of stationary phase used, *e.g.* spherical, irregular, fiber or porous. The q value is 2/3 for pellicular-type stationary phase materials and 1/30 for porous-type stationary phase materials.

The average residence time (t_d) of the sample molecules in the stationary phase is related to the thickness of the stationary phase (d_S) and the diffusion constant (D_S) of the sample molecules in the stationary phase. Thus $\sigma^2 = 2Dt$ (Equation 5.14) and $t_d = d_S^2/2D_S$ is obtained.

$R = V_M(V_M + KV_S)$ and $D_M = KV_S/V_0$, therefore $R = 1/(1 + D_M)$. Then, Equation 5.17 is converted into:

$$\sigma^2 = [qR(1 - R)uLd_S^2]/D_S$$
$$= (qD_Md_S^2uL)/(1 + D_M)^2D_S$$

Therefore, the band broadening due to diffusion of sample molecules in the stationary phase (H_S) is given by:

$$H_S = (qD_Md_S^2u)/(1 + D_M)^2D_S \tag{5.18}$$

From this equation, the H_S value is small in a high-efficiency column. Such a column has a stationary phase with a thin film or small d_S value. That is, a high-performance separation can be achieved by using a stationary phase in which the depth into which the sample molecules can move is diminished. This means that a thinner or shallower stationary phase gives higher performance in reversed-phase liquid chromatography, and a smaller particle size gives higher performance in ion-exchange and size-exclusion liquid chromatography. However, speeding up the diffusion counteracts lowering the flow rate. For diffusion in the mobile phase, we can consider the eddy diffusion (H_F) of Equation 5.13, longitudinal diffusion (H_L) of Equation 5.15, and the diffusion toward the wall from the centre of a column (or space between particles) (H_D). The time taken for the sample molecules to move from the column side wall or particle surface is given by Equation 5.14, where the distance is given as $\omega_\alpha d_p$:

$$t = \omega_\alpha^2 d_p^2/2D_M$$

where ω_α is the ratio between the distance moved to the wall and the particle diameter (d_p). When ω_b is defined as the ratio between limited partial linear velocity and the average linear velocity, the distance for sample molecules movement (l) is $\omega_\beta ut$. Therefore, the standard deviation, σ, is:

$$\sigma_D^2 = l^2n = [(\omega_\beta ut)^2L]/ut = (\omega_\alpha^2\omega_\beta^2d_p^2uL)/2D_M$$
$$= (\omega d_p^2uL)/D_M$$

where $\omega = \omega_\alpha^2\omega_\beta^2/2$. Then the diffusion effect due to the mobile phase, H_D, is:

$$H_D = (\omega d_p^2u)/D_M \tag{5.19}$$

where ω is the column constant, which depends on the shape of the column and the packing conditions.

As measured, the H_S term means that a smaller particle size reduces the distance between particles, and thus the spreading due to diffusion of sample molecules is minimized. From Equations 5.18 and 5.19, a decrease in H value is achieved by increasing the diffusion speed (elevating the column temperature, using a low-density, low-viscosity solvent), decreasing the flow rate, reducing the thickness of the stationary phase, and decreasing the particle size.

The overall effects of peak diffusion can be summarized as follows:

$$H_F = 2\lambda d_p \tag{5.13}$$

$$H_L = (2\gamma D_M)/u \tag{5.15}$$

$$H_S = (q D_M d_s^2 u)/(1 + D_M)^2 D_S \tag{5.18}$$

$$H_D = (\omega d_p^2 u)/D_M \tag{5.19}$$

$$\begin{aligned} H &= H_F + H_L + H_S + H_D \\ &= 2\lambda d_p + (2\gamma D_M)/u + (q D_M d_s^2 u)/(1 + D_M)^2 D_S + (\omega d_p^2 u)/D_M \end{aligned} \tag{5.20}$$

Overall, the most effective factor in Equation 5.20 is the particle size. The smaller the particle size, the higher the column efficiency. Equations 5.13, 5.15, and 5.18 are depicted in Figure 5.6 against flow velocity as A, B, and C, respectively. The band spreading is thus influenced by Equation 5.15 at a low flow rate. The band spreading is influenced by Equations 5.18 and 5.19 at a high flow rate. For gas chromatography curve D is obtained.

In liquid chromatography, the diffusion rates are slower than that in gas chromatography, and the values of D_M and D_S are very small; therefore, the minimum H value is obtained at a low flow rate, as shown by curve E in Figure 5.6. The value of H increases slowly at higher flow rates in liquid chromatography. An experimental result is shown in Figure 5.7. The HETP was minimal at a certain flow rate, and the measured optimum value was less than 10 μm for this column. The optimum flow rate was about 0.9 ml min^{-1}, corresponding to a linear flow velocity of about 55 mm min^{-1}.

Experimentally, high-performance separations can be performed at lower flow rates using small particle-size stationary phase materials as shown in Figure 1.1. A fast analysis can be achieved at a higher flow rate when the H

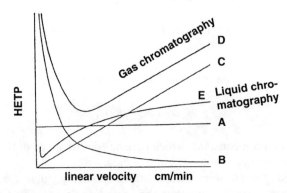

Figure 5.6   *Van Deemter curves relating H to linear flow velocity. A, eddy diffusion term; B, molecular diffusion term; C, resistance to mass transfer term; D, summation for gas chromatography; and E, summation for liquid chromatography.*

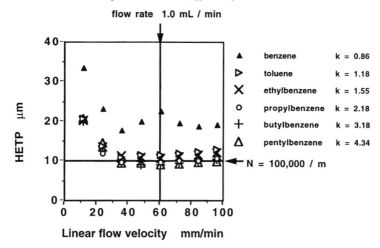

Figure 5.7 *HETP of compounds with different k values related to linear flow velocity. Column: 5 μm ODS-bonded silica gel, 15 cm × 4.6 mm i.d.; eluent: 80% aqueous acetonitrile.*

value is smaller; however, Equations 5.18 and 5.19 indicate that a high flow rate operation has negative effect. Curve E in Figure 5.6 indicates that a high flow rate operation does not significantly increase the *H* value in liquid chromatography. This means that a fast, efficient analysis can be performed by using a short column packed with small particle size stationary phase materials.

The pressure drop, Δp, inside a column is given by Equation 5.21:

$$\Delta p = (u\eta Lf)/\kappa^{\circ} \tag{5.21}$$

where η is the viscosity of the eluent, L is the column length, u is the linear flow velocity, f is the sum of the void volume of a column, and κ° is the parameter indicating the column permeability. When the void volume between the stationary phase materials is ε, κ° is given by Equation 5.22:

$$\kappa^{\circ} = \frac{d_p^2}{180} \times \frac{\varepsilon^3}{(1 - \varepsilon)^2} \tag{5.22}$$

From $t_R = L(D_M + 1)/u$ and Equation 5.21, the following equation is obtained:

$$t_R = [\eta L^2 f(1 + D_M)]/\Delta p \kappa^{\circ} \tag{5.23}$$

Thus when η, f, and L are constant, shortening the analysis time can be achieved by increasing the pressure drop and the κ° value. Increasing κ° can be done by decreasing ε and/or increasing the particle size, d_p, of the stationary phase material. Increasing the particle size decreases the resolution, R_s; therefore, small particle-size stationary phase materials must be densely packed. As a result, a high-pressure operation is unavoidable.

In conclusion, the smaller is the particle size the better the separation; however, a high pressure drop requires the optimum design of columns and instruments. A column well packed with spherical small particles has a high plate number, and is suitable for the separation of a homologous series of compounds. However, the separation of isomers requires a high selectivity of stationary and mobile phases as described in Chapters 3 and 4.

3   Bibliography

See Chapter 1 for details.

CHAPTER 6

Influence of Physical Chemistry on Separations in Liquid Chromatography

1 Quantitative Structure–Retention Relationships in Reversed-phase Liquid Chromatography

Various methods have been employed for the prediction of retention times in reversed-phase liquid chromatography.

As the alkyl chain length of fatty acids increases, their solubility in water becomes poorer and that in *n*-hexane becomes higher. Such a phenomenon is easily understood from a comparison of the properties of acetic acid and butter fat. In reversed-phase liquid chromatography, the retention time of fatty acids is related to the number of methylene units, *i.e.* the chain length. This relationship is a basic concept used to study retention times in reversed-phase liquid chromatography. In early work the number of carbon atoms of the aliphatic chain was related to the retention of alcohols[13] and phenols.[14–16] Even the polymerization of very polar compounds, such as amino acids and monosaccharides, makes them less soluble in water than their monomers. This means that polymerized molecules become more hydrophobic and their retention times become longer in reversed-phase liquid chromatography. Therefore, the retention time of a homologous series of compounds is easily predicted from the number of monomer units, and sometimes by the carbon numbers. It can also be related to molecular size, van der Waals volume, and surface area.

A quantitative analysis of the structure–retention relationship can be derived by using the relative solubility of solutes in water. One parameter is the partition coefficient, $\log P$, of the analyte measured as the octanol–water partition distribution. In early work, reversed-phase liquid chromatography was used to measure $\log P$ values for drug design. $\log P$ values were later used to predict the retention times in reversed-phase liquid chromatography. The calculation of the molecular properties can be performed with the aid of computational chemical calculations. In this chapter, examples of these quantitative structure–retention relationships are described.

Prediction of Retention Times from log *P* in Reversed-phase Liquid Chromatography

Prediction of Partition Coefficient (log P)

The measurement of the solubility of drugs in polar and non-polar media is very important in the pharmaceutical field. One method proposed to describe this solubility is the partition coefficient between octanol and water. The mathematical calculation of an octanol–water partition coefficient from values for functional groups was first proposed by Hansch *et al.* as Hansch's π constants,[1] and was later developed by Rekker as hydrophobic fragmental constants (log *P*).[2] This method was further improved by the use of molecular connectivities.[17] The prediction of log *P* values can be performed by either a computer program or by manual calculation. For example, approximate partition coefficients (log *P*) have been calculated by Rekker's method:

$$\log P = \sum f_i \qquad (6.1)$$

A part of his fragmental constant tables is given in Table 6.1. (The predicted log *P* values are summarized in Table 6.4.)

Attempts have also been made to determine octanol–water partition coefficients empirically using liquid chromatography by comparison of retention times with those of compounds of known log *P*.[3–12]

Correlation between log P and log k Values

It seemed possible to correlate the elution order in reversed-phase liquid chromatography with the octanol–water partition coefficient, log *P*. For example, the partition coefficients calculated by Rekker's method showed a

Table 6.1 *Rekker's hydrophobic fragment constants*

Fragment	Δ log P	Fragment	Δ log P	Fragment	Δ log P
C_6H_5	1.886	(al)COOH	−0.954	(al)OCH$_2$COOH	−1.155
C_6H_4	1.688	(ar)COOH	−0.093	(ar)OCH$_2$COOH	−0.581
C_6H_3	1.431	(al)COO	−1.292		
CH$_3$	0.702	(ar)COO	−0.431		
CH$_2$	0.530	(al)CO	−1.703		
CH	0.235	(ar)CO	−0.842		
CH (ar)	0.36	(al)O	−1.581		
C (al)	0.15	(ar)O	−0.433		
CH$_2$=CH$_2$	0.935	(al)OH	−1.491		
CH=CH	0.73	(ar)OH	−0.343		
H	0.175	(ar)COH	−0.38		

ar: substituent on aromatic compound, al: substituent on aliphatic compound; hydrophobic fragment constants in octanol–water from Rekker[2].

linear relation to $\log k$ values measured in reversed-phase liquid chromatography:[18]

$$\log k = y \times \log P + m \tag{6.2}$$

where y and m are constants in a given system. An example is given in Figure 6.1. The values of $\log P$ obtained from Equation 6.1 using the measured $\log k$ of 58 compounds, including alkanols, alkylphenones, alkyl benzoates, polycyclic aromatic hydrocarbons, halogenated benzenes, *etc.*, were related to the $\log P$ values calculated according to Rekker's method. The average difference was within $0.16 \log P$ unit. The correlation coefficient was 0.975.[19] However, the peak for aniline showed tailing. An organic modifier effect on $\log P$ was described, and a method for the qualitative analysis of free fatty acids was proposed.[13,18,20]

Prediction of Retention Times from log P

The prediction of retention times in a given eluent from $\log P$ has been proposed for aromatic hydrocarbons.[19] The $\log k$ values of phenols[21] and nitrogen-containing compounds[22] were also related to their $\log P$, and the calculated $\log P$ was used for the qualitative analysis of urinary aromatic acids, *i.e.* for the identification of metabolites in urine from the differences of $\log P$ in reversed-phase liquid chromatography.[23,24]

A good correlation was obtained in 20–80% acetonitrile–water mixtures. The standard non-ionic compounds used to evaluate the columns were 2-hydroxy-acetophenone, coumarin, acetophenone, indole, propiophenone, butyro-phenone, isopropyl benzoate, butyl benzoate, and isopentyl benzoate. The plotted lines for the linear relationship measured in five different proportions

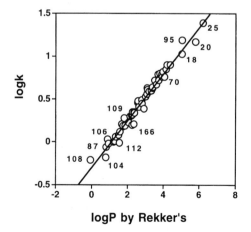

Figure 6.1 *Relationship of Rekker's log P values to log k values on octadecyl-bonded silica gel column at 30 °C. Column, Chromosorb LC7; eluent, 60% aqueous acetonitrile. Number beside symbols: see Table 6.4.*

of acetonitrile–water mixtures merged at a single point. These slopes (y) have a fourth-degree polynomial relation with the acetonitrile concentration (x) (Equation 6.3, where a_i is a contant):

$$y = \sum_{i=0}^{i=n} a_i \times x^i \qquad n = 2\text{–}8 \tag{6.3}$$

Further, an optimization of the organic modifier concentration for the separation of known compounds was proposed. When the value of the actual column plate number, the resolution, and log P of pairs of compounds a and b were known, the percentage concentration (x) required for their separation was calculated from Equations 6.4 to 6.6:[19,20]

$$R_s = [(\alpha - 1)/\alpha]0.25\sqrt{N_{\text{eff}}} \tag{6.4}$$

$$y = [\log \sqrt{N_{\text{eff}}} - \log(\sqrt{N_{\text{eff}}} - 4R_s)]/[\log P_a - \log P_b] \tag{6.5}$$

$$x = \sum_{j=0}^{j=n} b_j \times y^i \tag{6.6}$$

The difference between the log k values measured and those predicted by Equations 6.1 and 6.2 from log P values calculated by Rekker's hydrophobic fragmental constant was within 5%. An example of log k values obtained experimentally compared with those predicted from Rekker's log P values is shown in Figure 6.2, where the predicted and measured retention factors of aromatic acids are shown. The correlation coefficient was 0.993 ($n = 32$), and the slope was 1.014. The precision is very high.

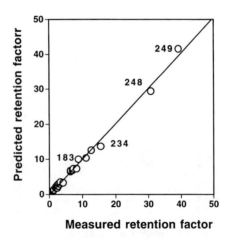

Figure 6.2 *Comparison of measured and predicted retention factors of aromatic acids from Rekker's log P values. Column, 10 μm polystyrene gel, Hitachi 3011, 15 cm × 4.6 mm i.d.; eluent, 30% aqueous acetonitrile containing 50 mM phosphoric acid at 55 °C. Numbers beside symbols: see Table 6.4.*

Prediction of Retention Time of Ionic Compounds from log P and pK_a

The retention times of neutral non-ionized compounds can be predicted by the above calculation method, but those of ionized compounds are also very important. The inclusion of the dissociation constant in the calculation made it possible to predict the retention times of ionized aromatic acids.[25,26] The dissociation constant was calculated by the method proposed by Perrin *et al.*[27]

When a compound is ionized, the retention becomes shorter than that of the molecular form, and the retention factor of the ionized form is obtained from Equation 6.7:

$$k = \frac{k_m + k_i(K_a/[H^+])}{1 + K_a/[H^+]} \tag{6.7}$$

where k_m is the retention factor of the non-ionized acid and can be obtained from log P as above and k_i is the retention factor of the 100% ionized acid. The k_i value cannot be predicted mathematically at present, but the value is close to zero in many cases.

Equation 6.7 was further modified to improve the precision in low pH solutions where the strong acid used for pH control reduced the retention of weak acids by excluding them from the hydrophobic surface of the stationary phase.[26] A modified version of Equation 6.7 is given in the following form (a slight modification of constant A improved the precision):

$$k = A\frac{k_m - k_i}{2}\tanh(pK_a - pH) + \frac{k_m + k_i}{2} \tag{6.8}$$

An example of the relationship between the predicted and observed log k values at pH 4.00 is shown in Figure 6.3.

The agreement between the observed and predicted k values of aromatic acids was within 10%. The correlation coefficient was 0.954 ($n = 32$). An error of greater than 10% for 3-hydroxy-2-naphthoic acid and 2-hydroxybenzoic acid was attributed mainly to an error in their K_a values.[25] The partition coefficient, log P, and dissociation constant, pK_a, of analytes can be obtained by simple calculations and by computational chemical calculations, and thus the retention time can be predicted in reversed-phase liquid chromatography.

Calculation of Dissociation Constant pK_a from Hammet's Equation

The dissociation constant of an analyte can be calculated mathematically from Hammet's equation.[27] The organic solvent effect on the pK_a has also been examined:[26]

$$pK_a = A + B\sum\sigma_i$$

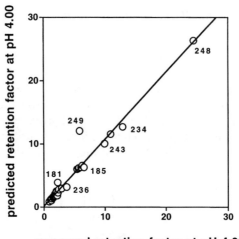

measured retention factor at pH 4.00

Figure 6.3 *Comparison of measured and predicted retention factors of aromatic acids from Rekker's log P values. Column, 5 µm polystyrene gel, Hitachi 3013, 15 cm × 4.6 mm i.d.; eluent, 20% aqueous acetonitrile containing 50 mM sodium phosphate at 55 °C. Numbers beside symbols: see Table 6.4.*

where A and B are constants for individual groups of compounds, and σ is Hammet's σ constant. However, it is difficult to predict the pK_a values of *ortho*-substituted compounds. Examples of pK_as from modified Hammet's equations are:[25]

Benzoic acids:	$4.20-1.00 \sum \sigma$
2-Hydroxybenzoic acids:	$4.20-1.13 \sum \sigma$
Phenylacetic acids:	$4.30-0.35 \sum \sigma$
Cinnamic acids:	$4.38-0.66 \sum \sigma$
Mandelic acids:	$3.38-0.454 \sum \sigma$

The original equations for phenolic and nitrogen-containing compounds are:[27]

Phenolic compounds:	$9.92-2.23 \sum \sigma$
Anilines	$4.58-2.28 \sum \sigma$
Pyridines (1):	$5.25-5.90 \sum \sigma$
Pyridines (2):	$5.39-5.70 \sum \sigma$
Quinolines (2-substituted)	$5.12-9.04 \sum \sigma$
Quinolines (8-substituted)	$4.64-3.11 \sum \sigma$
α-Naphthylamines:	$3.85-2.81 \sum \sigma$
β-Naphthylamines:	$4.29-2.81 \sum \sigma$
Benzylamines (ring substituted):	$9.39-1.05 \sum \sigma$
N-Substituted anilines	$5.06-3.46 \sum \sigma$
1-Aminoanthracene:	aniline $+ 0.17$
Quinoline:	pyridine (1) $+ 0.06$

where σ = Hammet's sigma constant. Examples of σ constants are given in

Table 6.2 *Hammet's substituent constants*

Substituent	σ_{meta}	σ_{para}	Substituent	σ_{meta}	σ_{para}
Br	0.39	0.22	CH_3	−0.06	−0.14
Cl	0.37	0.24	OCH_3	0.11	−0.28
H	0.00	0.00	NH_2	0.00	−0.57
I	0.35	0.15	CO_2H	0.35	0.44
OH	0.13	−0.38	C_2H_5	−0.07	−0.15

From Ref. 27.

Table 6.2. The calculated values are summarized in Table 6.4 (along with reference values).

2 Van der Waals Volume as the Basic Property

In a reversed-phase liquid chromatography system, if the hydrogen-bonding and Coulombic forces are negligible, the retention of molecules depends upon their size, and the presence of π electrons enhances the selectivity. The molecular size, *i.e.* the van der Waals volume, can be calculated by hand, by Bondi's method,[29] or by certain computer programs.

The possibility for predicting the retention time of polycyclic aromatic hydrocarbons from a combination of the calculated van der Waals volume and the reference values of the delocalization energy has been demonstrated.[29] The precision of a method for predicting the retention times of polycyclic aromatic hydrocarbons in reversed-phase liquid chromatography has been improved by a combination of the van der Waals volume calculated by the computer program MOPAC-BlogP and the kinetic energy calculated by molecular dynamics.[30] However, the precision for polychlorinated benzenes could not be improved by this method.[31] The prediction of retention times of phenols was further improved by using the ionization potential calculated by the computational chemical program CAChe[TM].

Calculation of van der Waals Volume

Bondi's calculation method is simple. The van der Waals volumes are the sum of the van der Waals volumes of fragments, as given in Table 6.3. The calculated van der Waals volumes are summarized in Table 6.4. However, these volumes are different from those calculated using the MOPAC-BlogP program, even though the correlation is excellent.

Prediction of Retention Times Based on van der Waals Volumes

The relationship between the $\log k$ and the van der Waals volumes of alkanes, alkylbenzenes, and polycyclic aromatic hydrocarbons measured in 70% aqueous acetonitrile on octadecyl-bonded silica gels[31] is shown in Figure 6.4.

Table 6.3 *Bondi's group contributions to the van der Waals volume of hydrocarbons*

Groups	VWV^a	Groups	VWV^a
Alkane		*Olefinic*	
C	3.33	$=C=$	6.96
CH	6.78		
CH$_2$	10.23	$\diagdown C=C \diagup$	10.02
CH$_3$	13.67		
CH$_4$	17.12	$=CH-$	8.47
		$=CH_2$	11.94
Aromatic		$\diagdown C=CH_2$	16.95
$\diagdown C-$(cond.)	4.74	$\diagdown C=CH-$	13.49
$\diagdown C-R$	5.54	*Acetylenic*	
$\diagdown C-H$	8.05	$-C\equiv$	8.05
Benzene	48.36	$\equiv C-H$	11.55
Phenyl	45.84	$\equiv C-$(diacetylene)	7.82

^{a} Van der Waals volume (cm^3 mol^{-1}) from Ref. 28.

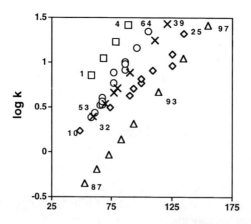

van der Waals volume by Bondi   cm^3/mole

Figure 6.4   *Relationship of log k values measured on ODS-bonded silica gel to Bondi's van der Waals volumes. Column, Develosil ODS, 15 cm × 4.6 mm i.d.; eluent, 70% aqueous acetonitrile at 30 °C. Numbers beside symbols: see Table 6.4:* $\diamond$, *Polycyclic aromatic hydrocarbons;* ×, *alkylbenzenes;* ○, *halogenated benzenes;* △, *alkanols; and* □, *alkanes.*

Table 6.4 *Molecular properties of non-ionic compounds*

No.	Chemical	M_r	VWV^a	SA^b	IP^c	$B\log P^d$	$-\Delta H^e$	VWV^f	$\log P^g$	$-\Delta H^h$
1	Pentane	72.150	95.970	136.780	11.060	3.395	5175	58.03	2.994	2.301
2	Hexane	86.177	112.787	158.420	11.060	3.972	6048	68.26	3.524	2.367
3	Heptane	100.203	129.552	180.714	11.060	4.570	6919	78.49	4.054	2.957
4	Octane	114.230	146.408	202.272	11.060	5.139	7794	88.72	4.584	3.372
5	Nonane	128.257	163.232	224.205	11.064	5.698	8668	—	—	—
6	Decane	142.284	180.007	246.467	11.062	6.267	9539	109.18	5.644	—
7	Undecane	156.311	196.800	267.965	11.063	6.738	10411	—	—	—
8	Dodecane	170.337	213.653	289.660	11.063	7.209	11286	129.64	6.704	—
9	Tridecane	184.364	230.555	311.976	11.032	7.648	12164	—	—	—
10	Benzene	78.113	83.789	109.752	9.652	2.432	3402	43.86	2.280	1.484
11	Naphthalene	128.173	127.598	156.999	8.711	3.729	4897	73.96	3.208	1.926
12	Acenaphthylene	152.195	144.969	173.766	8.943	4.274	5359	—	—	—
13	Acenaphthene	154.211	150.675	181.904	8.492	4.297	5668	—	—	—
14	Biphenyl	154.211	155.227	189.990	8.950	4.316	6237	90.08	3.772	2.034
15	Fluorene	166.222	161.639	194.923	8.717	4.545	6145	93.22	3.906	2.235
16	Phenanthrene	178.233	171.286	201.409	8.617	4.799	6519	99.56	4.378	2.476
17	Anthracene	178.233	171.486	204.540	8.123	4.808	6539	99.56	4.378	2.543
18	Pyrene	202.255	187.603	216.623	8.130	5.246	6847	109.04	5.030	2.885
19	Fluorancene	202.255	188.723	220.619	8.629	5.248	6991	—	—	—
20	Chrysene	228.293	214.558	244.822	8.371	5.581	8201	125.16	5.545	3.207
21	Tetracene	228.293	215.352	252.509	7.747	5.634	8733	125.16	5.790	3.442
22	Benzo[a]anthracene	228.293	215.841	251.956	7.938	5.554	8470	—	—	—
23	p-Terphenyl	230.309	226.657	273.024	8.746	5.710	9205	—	—	—
24	Benzo[e]pyrene	252.315	230.588	256.615	8.214	5.899	8595	—	—	—
25	Benzopyrene	252.315	231.048	260.079	7.918	5.912	8571	134.64	6.200	3.700
26	Benzo[k]fluoranthene	252.315	232.418	268.140	8.302	5.962	8669	—	—	—
27	Benzo[b]fluorancene	252.315	232.462	267.348	8.559	5.941	8663	—	—	—

(*continued*)

Table 6.4 *Continued*

No.	Chemical	M_r	VWV^a	SA^b	IP^c	$B log P^d$	$-\Delta H^e$	VWV^f	$log P^g$	$-\Delta H^h$
28	Benzo[ghi]perylene	276.337	247.198	272.981	8.032	6.188	8900	—	—	—
29	Dibenzo[ah]anthracene	278.353	259.435	296.969	8.030	6.061	10099	—	—	—
30	Pentacene	278.353	259.277	299.886	7.491	6.197	9885	150.76	6.960	—
31	Indeno[1,2,3-cd]pyrene	276.337	248.683	279.958	8.138	6.197	9041	—	—	—
32	Toluene	92.140	100.456	131.549	9.333	2.778	4510	59.51	2.588	1.658
33	Ethylbenzene	106.167	117.067	152.079	9.298	3.189	5217	69.74	3.118	1.794
34	Isopropylbenzene	120.194	134.032	172.293	9.388	3.614	6159	77.02	3.525	1.906
35	Propylbenzene	120.194	133.897	173.691	9.297	3.613	6107	79.97	3.648	2.053
36	Butylbenzene	134.221	150.713	195.232	9.298	4.052	7002	90.20	4.178	2.331
37	Pentylbenzene	148.247	167.564	217.077	9.301	4.496	7847	—	—	—
38	Hexylbenzene	162.274	184.315	239.063	9.317	4.933	8715	110.66	5.238	2.797
39	Heptylbenzene	176.301	201.171	260.952	9.300	5.357	9580	120.89	5.758	3.200
40	Octylbenzene	190.328	217.884	283.590	9.299	5.764	10458	131.12	6.297	3.673
41	Nonylbenzene	204.355	234.706	305.218	9.299	6.119	11305	141.35	3.453	6.828
42	Decylbenzene	218.381	251.540	326.927	9.301	6.433	12116	151.58	3.655	—
43	Acetophenone	120.151	119.403	153.029	9.934	1.936	5546	—	1.75	—
44	Propiophenone	134.177	136.200	174.031	9.925	2.346	6453	—	2.28	—
45	Butyrophenone	148.204	153.052	196.058	9.925	2.768	7313	—	2.81	—
46	Pentanophenone	162.231	169.817	217.968	9.922	3.197	8153	—	—	—
47	Hexanophenone	176.258	186.582	240.137	9.926	3.625	9035	—	—	—
48	Heptanophenone	190.285	203.415	261.523	9.924	4.009	9915	—	—	—
49	Octanophenone	204.311	220.180	283.668	9.924	4.393	10772	—	—	—
50	Nonanophenone	218.388	236.945	305.856	9.924	4.751	11640	—	—	—
51	Decanophenone	232.365	253.710	327.255	9.924	5.035	12507	—	—	—
52	Undecanophenone	246.392	270.484	349.826	9.925	5.344	13371	—	—	—
53	Chlorobenzene	112.559	97.300	126.293	9.560	2.826	4047	57.84	2.808	1.656
54	1,2-Dichlorobenzene	147.004	110.741	141.224	9.602	3.295	4744	67.32	3.532	1.884
55	1,3-Dichlorobenzene	147.004	110.911	142.517	9.686	3.324	4744	67.32	3.532	1.950
56	1,4-Dichlorobenzene	147.004	110.844	142.627	9.523	3.393	4745	67.32	3.532	1.944

57	1,2,3-Trichlorobenzene	181.449	124.174	156.074	9.785	3.771	5451	76.80	4.197	2.095
58	1,2,4-Trichlorobenzene	181.449	124.307	157.548	9.653	3.828	5466	76.80	4.197	2.163
59	1,3,5-Trichlorobenzene	181.449	124.505	158.784	9.921	3.889	5480	76.80	4.197	2.189
60	1,2,3,4-Tetrachlorobenzene	215.894	137.542	170.685	9.734	4.256	6180	86.28	4.944	2.351
61	1,2,3,5-Tetrachlorobenzene	215.894	137.770	172.135	9.763	4.305	6208	86.28	4.944	2.349
62	1,2,4,5-Tetrachlorobenzene	215.894	137.730	172.368	9.656	4.352	6209	86.28	4.944	2.351
63	Pentachlorobenzene	250.339	150.870	185.175	9.786	4.736	6938	95.76	5.691	2.542
64	Hexachlorobenzene	284.784	164.037	197.770	9.911	5.176	7667	105.24	6.438	2.741
65	2-Chlorotoluene	126.585	114.010	146.978	9.415	3.236	5079	—	3.40	—
66	3-Chlorotoluene	126.585	114.119	148.096	9.427	3.211	5217	—	3.40	—
67	4-Chlorotoluene	126.585	114.140	148.082	9.297	3.206	5217	—	3.40	—
68	2-Bromotoluene	171.036	121.894	153.533	9.473	3.606	5221	—	3.61	—
69	4-Iodotoluene	218.037	129.514	162.180	9.410	3.918	5432	—	3.93	—
70	1-Chloro-4-iodobenzene	238.455	126.392	156.666	9.621	4.110	5005	—	4.06	—
71	1-Bromo-3-chlorobenzene	191.455	118.912	149.925	9.714	3.693	4915	—	3.74	—
72	2-Bromo-4-chloroacetophenone	233.492	154.618	194.224	9.914	3.232	7367	—	2.57	—
73	Bromobenzene	157.010	105.336	133.720	9.604	3.217	4208	—	3.017	—
74	Iodobenzene	204.010	112.716	140.247	9.645	3.590	4292	—	3.334	—
75	Methyl benzoate	136.150	128.120	164.986	10.020	1.969	6205	—	2.15	—
76	Isopropyl benzoate	164.204	161.823	206.896	9.977	2.776	7976	—	3.09	—
77	Butyl benzoate	178.230	178.537	229.265	9.997	3.228	8719	—	3.74	—
78	Isopentyl benzoate	192.257	195.268	248.769	9.991	3.561	9603	—	4.15	—
79	p-Xylene	106.167	117.182	153.244	9.059	3.224	5684	—	3.10	—
80	1,4-Dimethoxybenzene	138.170	134.452	173.357	8.869	1.804	6645	—	2.23	—
81	Phenethyl acetate	150.177	144.920	187.287	9.493	2.377	7293	—	2.36	—
82	Benzyl ethyl ether	122.166	125.900	163.363	8.961	2.190	5910	—	2.64	—
83	2-Methylnaphthalene	142.200	144.195	178.866	8.617	4.068	6013	—	3.70	—
84	2,4-Dimethylnaphthalene	156.227	160.782	198.828	8.503	4.429	7056	—	4.20	—
85	1-Chloronaphthalene	162.618	140.961	171.558	8.781	4.105	5619	—	3.80	—
86	Propanol	60.096	70.540	104.025	10.847	0.362	4222	—	0.271	—
87	Butanol	74.122	87.399	125.838	10.845	0.922	5042	52.40	0.801	0.833

(continued)

Table 6.4 *Continued*

No.	Chemical	M_r	VWV^a	SA^b	IP^c	$B \log P^d$	$-\Delta H^e$	VWV^f	$\log P^g$	$-\Delta H^h$
88	Pentanol	88.149	104.174	147.423	10.851	1.497	5911	62.63	1.331	0.957
89	Hexanol	102.176	121.084	169.712	10.846	2.093	6786	72.86	1.861	1.214
90	Heptanol	116.203	137.862	191.712	10.851	2.670	7661	83.09	2.391	1.568
91	Octanol	130.230	154.627	213.447	10.849	3.259	8535	93.32	2.921	1.888
92	Nonanol	144.256	171.395	234.930	10.851	3.791	9406	—	—	—
93	Decanol	158.283	188.472	257.551	10.845	4.347	10279	113.78	3.981	2.508
94	Undecanol	172.310	204.994	278.645	10.852	4.800	11157	—	—	—
95	Dodecanol	186.337	222.059	301.238	10.847	5.280	12020	134.24	5.041	3.249
96	Tridecanol	200.364	238.703	322.333	10.853	5.634	12880	—	—	—
97	Tetradecanol	214.390	255.670	345.123	10.848	6.046	13760	154.70	6.701	—
98	Pentadecanol	228.417	272.009	365.820	10.852	6.294	14636	—	—	—
99	Hexadecanol	242.444	289.335	389.083	10.846	6.603	15515	175.16	7.101	—
100	Benzyl alcohol	108.140	108.505	141.102	9.362	1.848	5145	—	0.93	—
101	Cinnamanol	120.151	119.356	153.475	8.588	2.038	5769	—	1.49	—
102	Indole	117.150	112.880	142.179	8.405	1.602	4705	—	2.06	—
103	Coumarin	146.145	127.931	158.467	9.461	1.665	5403	—	1.57	—
104	Benzamide	121.138	114.245	148.148	9.948	1.212	5462	—	0.78	—
105	Salicylamide	137.138	121.914	155.779	9.509	0.718	6029	—	1.28	—
106	Phenacetin	179.218	173.529	222.040	8.348	1.566	5403	—	0.88	—
107	Uracil	112.088	90.666	121.421	9.972	-0.588	4565	—	—	—
108	Caffeine	194.193	167.556	210.514	8.964	0.326	9068	—	-0.07	—
109	Nitrobenzene	123.111	106.298	137.585	10.557	1.987	4876	—	1.81	—
110	1-Fluoro-2,4-dinitrobenzene	186.099	132.899	169.714	11.264	1.685	6945	—	1.67	—
111	Benzyl methyl ketone	120.151	119.549	153.022	9.938	1.938	5518	—	1.75	—

^{a} van der Waals volume ($\mathring{A}^3$ mol^{-1}); ^{b} surface area ($\mathring{A}^2$) and ^{d} partition coefficient are calculated by MOPAC-BlogP from CAChe^{TM}; ^{c} ionization potential (eV) and ^{e} enthalpy are calculated by MOPAC from CAChe^{TM}; ^{f} van der Waals volume calculated by Bondi's method; ^{g} partition coefficient calculated by Rekker's method; ^{h} enthalpy measured by liquid chromatography.

Table 6.4B *Molecular properties of acidic compounds*

No.	Chemical	M_r	VWV^a	SA^b	IP^c	$B \log P^d$	$-\Delta H^e$	VWV^f	$\log P^g$	$-\Delta H^h$	pK_a^i
112	Phenol	94.113	91.738	120.577	9.114	1.295	4115	53.88	1.54	1.980	10.02
113	2-Methylphenol	108.140	108.327	140.162	8.959	1.699	5157	65.03	2.05	1.683	10.32
114	3-Methylphenol	108.140	108.432	142.265	9.041	1.706	5238	65.03	2.05	—	10.09
115	4-Methylphenol	108.140	108.447	142.192	8.881	1.703	5304	65.03	2.05	1.910	10.27
116	2,3-Dimethylphenol	122.166	124.810	160.190	8.924	2.163	6237	76.18	2.49	1.926	10.54
117	2,4-Dimethylphenol	122.166	125.003	162.149	8.781	2.162	6298	76.18	2.49	1.872	10.60
118	2,5-Dimethylphenol	122.166	125.035	162.359	8.896	2.141	6312	76.18	2.49	1.880	10.41
119	2,6-Dimethylphenol	122.166	124.816	159.997	8.887	2.138	6316	76.18	2.49	1.941	10.63
120	3,4-Dimethylphenol	122.166	125.229	161.752	8.803	2.146	9186	76.18	2.49	1.906	10.36
121	3,5-Dimethylphenol	122.166	125.339	163.956	8.975	2.139	6442	76.18	2.49	1.889	10.19
122	2,3,4-Trimethylphenol	136.193	141.221	178.530	8.726	2.562	7158	—	—	—	—
123	2,3,5-Trimethylphenol	136.193	141.587	181.784	8.853	2.578	7307	87.33	3.02	1.947	—
124	2,3,6-Trimethylphenol	136.930	141.094	179.555	8.785	2.568	7222	87.33	3.02	1.940	—
125	2,4,6-Trimethylphenol	136.193	141.518	182.671	8.783	2.597	7433	87.22	3.02	1.898	10.88
126	2,3,4,5-Tetramethylphenol	150.220	157.430	197.433	8.633	2.921	8220	—	—	—	—
127	2,3,5,6-Tetramethylphenol	150.220	157.741	197.874	8.763	2.922	8330	98.48	3.55	1.960	—
128	Pentamethylphenol	164.247	173.376	212.220	8.609	3.205	9126	—	—	—	—
129	2-Ethylphenol	122.166	124.897	160.104	8.998	2.099	5922	75.26	2.58	1.877	10.2
130	3-Ethylphenol	122.166	124.979	162.093	9.017	2.096	5984	75.26	2.58	1.838	9.9
131	4-Ethylphenol	122.166	125.084	161.793	8.862	2.118	5975	75.26	2.58	1.886	10.0
132	2-Chlorophenol	128.558	105.235	136.109	9.259	1.743	4775	63.03	2.27	1.899	8.48
133	3-Chlorophenol	128.558	105.249	136.771	9.335	1.731	4808	63.03	2.27	2.118	9.02
134	4-Chlorophenol	128.558	105.220	137.032	9.125	1.773	4810	63.03	2.27	2.085	9.38
135	2,3-Dichlorophenol	163.003	118.717	150.720	9.430	2.203	5560	72.51	2.93	2.061	7.45
136	2,4-Dichlorophenol	163.003	118.954	152.192	9.230	2.252	5521	72.51	2.93	2.029	7.89
137	2,5-Dichlorophenol	163.003	118.947	152.222	9.303	2.304	5512	72.51	2.93	2.031	7.50
138	2,6-Dichlorophenol	163.003	118.595	149.937	9.374	2.278	5548	72.51	2.93	2.032	6.79

(continued)

Table 6.4B *Continued*

No.	Chemical	M_r	VWV^a	SA^b	IP^c	$B \log P^d$	$-\Delta H^e$	VWV^f	$\log P^g$	$-\Delta H^h$	pK_a^i
139	3,4-Dichlorophenol	163.003	118.826	152.197	9.280	2.221	5514	72.51	2.93	2.133	8.39
140	3,5-Dichlorophenol	163.003	118.965	153.141	9.538	2.276	5548	72.51	2.93	2.209	8.18
141	2,3,4-Trichlorophenol	197.448	132.121	165.377	9.380	2.684	6244	81.99	3.68	2.083	7.59
142	2,3,5-Trichlorophenol	197.448	133.301	167.060	9.484	2.740	6251	81.99	3.68	2.374	7.23
143	2,3,6-Trichlorophenol	197.448	131.961	164.831	9.535	2.735	6293	81.99	3.68	2.093	6.12
144	2,4,5-Trichlorophenol	197.448	132.324	167.370	9.322	2.775	6282	81.99	3.68	2.157	7.33
145	2,4,6-Trichlorophenol	197.448	132.193	166.307	9.390	2.802	6286	81.99	3.68	—	6.42
146	3,4,5-Trichlorophenol	197.448	132.273	166.399	9.422	2.722	6243	81.99	3.68	2.335	7.74
147	2,3,4,5-Tetrachlorophenol	231.893	145.518	179.888	9.457	3.170	6974	91.47	4.43	2.470	6.96
148	2,3,5,6-Tetrachlorophenol	231.893	145.328	179.604	9.629	3.238	6978	91.47	4.43	2.251	5.44
149	Pentachlorophenol	266.338	158.497	192.012	9.574	3.640	7748	100.95	5.17	2.439	5.26
150	2-Chloro-5-methylphenol	142.585	121.556	154.549	9.127	2.140	5860	74.51	2.71	—	—
151	4-Chloro-2-methylphenol	142.585	121.654	155.526	9.023	2.176	5891	74.51	2.71	—	—
152	4-Chloro-3-methylphenol	142.585	122.025	157.213	9.031	2.189	5852	74.51	2.71	—	—
153	2-Bromophenol	173.009	112.895	141.882	9.301	2.197	4938	66.48	2.48		8.44
154	3-Bromophenol	173.009	113.321	144.179	9.348	2.229	4971	66.48	2.48	2.039	9.03
155	4-Bromophenol	173.009	113.358	144.363	9.190	2.148	4954	66.48	2.48	2.024	9.36
156	2,4-Dibromophenol	251.905	133.715	162.464	9.461	3.001	5852	79.08	3.35	2.021	7.8
157	2,6-Dibromophenol	251.905	134.337	164.099	9.444	2.956	5831	79.08	3.35		6.6
158	2-Nitrophenol	139.110	113.337	146.167	9.952	1.644	5640	67.11	1.92	—	7.23
159	3-Nitrophenol	139.110	114.417	148.348	9.948	1.571	5655	68.16	1.27	—	8.40
160	4-Nitrophenol	139.110	114.403	147.925	10.072	1.723	5584	68.16	1.27	—	7.15
161	2,4-Dinitrophenol	184.108	136.132	172.035	10.757	1.221	7105	81.39	1.58	—	4.09
162	2,5-Dinitrophenol	184.108	136.159	171.927	10.613	1.420	7105	81.39	1.58	—	5.22
163	2,6-Dinitrophenol	184.108	135.574	169.758	10.656	1.167	7172	81.39	1.58	—	3.71
164	3,4-Dinitrophenol	184.108	137.036	175.184	10.718	0.849	7167	82.44	0.93	—	5.43
165	2-Hydroxyacetophenone	136.150	126.970	161.167	9.319	1.461	6265	—	1.21	—	—
166	4-t-Butylphenol	150.220	158.057	198.996	8.893	2.887	7599	—	3.60	—	—
167	Propyl 4-hydroxybenzoate	180.203	169.667	219.136	9.516	2.488	8719	—	2.67	—	—

168	Butyl 4-hydroxybenzoate	194.230	186.310	240.792	9.520	2.930	9581	—	3.20	—	—
169	4-Chloro-3,5-dimethylphenol	156.612	138.449	177.464	8.892	2.624	6891	—	3.24	—	—
170	1,2-Dihydroxybenzene	110.112	99.793	130.523	8.796	0.878	5004	—	1.00	—	—
171	1,3-Dihydroxybenzene	110.112	99.888	131.163	9.044	0.853	4836	—	1.00	—	9.81
172	1,4-Dihydroxybenzene	110.112	99.852	131.027	8.736	0.878	4915	—	1.00	—	10.35
173	1-Hydroxynaphthalene	144.173	135.452	166.326	8.465	2.618	5637	—	2.86	—	—
174	2-Hydroxynaphthalene	144.173	135.586	167.716	8.643	2.612	5613	—	2.86	—	—
175	1-Hydroxy-2,4-dinitronaphthalene	234.168	178.767	212.816	9.792	2.173	8717	—	2.45	—	—
176	Benzoic acid	122.123	110.724	142.853	10.083	1.612	5159	65.36	1.94	2.378	4.200
177	3,4,5-Trihydroxybenzoic acid	170.121	134.292	171.289	9.372	0.756	7595	—	0.72	—	4.320
178	3,4-Dihydroxybenzoic acid	154.122	126.668	163.122	9.230	0.992	6773	—	0.99	—	4.450
179	3,5-Dihydroxybenzoic acid	154.122	126.699	164.064	9.407	0.968	6681	—	0.94	—	3.940
180	2,4-Dihydroxybenzoic acid	154.122	126.388	161.686	9.505	0.765	6652	—	1.30	—	3.250
181	4-Hydroxybenzoic acid	138.123	118.751	153.441	9.605	1.260	5874	—	1.28	—	4.580
182	3-Hydroxybenzoic acid	138.123	118.740	153.334	9.519	1.361	5916	—	1.37	—	4.070
183	2-Hydroxybenzoic acid	138.123	118.396	151.204	9.506	1.146	5967	—	2.18	—	2.821
184	4-Hydroxy-3-methoxybenzoic acid	168.149	143.959	184.413	9.100	1.501	7643	—	1.31	—	4.470
185	3-Methoxybenzoic acid	152.149	135.989	174.473	9.391	1.825	6799	—	1.99	—	4.090
186	4-Methoxybenzoic acid	152.149	136.002	174.655	9.479	1.688	6790	—	1.86	—	4.480
187	2-Methylbenzoic acid	136.150	126.868	160.859	9.728	2.038	6095	76.51	2.297	2.571	—
188	3-Methylbenzoic acid	136.150	127.432	164.572	9.747	2.020	6343	76.51	2.297	2.465	4.270
189	4-Methylbenzoic acid	136.150	127.336	164.514	9.824	2.014	6325	76.51	2.297	2.428	4.340
190	2-Ethylbenzoic acid	150.177	143.994	182.946	9.781	2.390	7112	86.74	—	—	—
191	3-Ethylbenzoic acid	150.177	143.930	184.475	9.707	2.422	7042	86.74	—	—	—
192	4-Ethylbenzoic acid	150.177	143.963	184.244	9.784	2.413	7042	86.74	2.827	2.539	4.350
193	2,4-Dimethylbenzoic acid	150.177	143.412	181.614	9.636	2.423	7255	87.66	—	2.623	—
194	2,5-Dimethylbenzoic acid	150.177	143.428	181.896	9.446	2.448	7254	87.66	—	2.623	—
195	2,6-Dimethylbenzoic acid	150.177	143.334	179.269	9.520	2.429	7115	87.66	—	2.273	—
196	3,4-Dimethylbenzoic acid	150.177	144.135	186.656	9.636	2.472	7294	87.66	—	2.486	4.440
197	3,5-Dimethylbenzoic acid	150.177	144.135	186.056	9.581	2.472	7518	87.66	—	2.513	4.340
198	2,4,6-Trimethylbenzoic acid	164.204	159.835	200.682	9.488	2.808	8298	98.81	—	2.393	—

(continued)

Table 6.4B *Continued*

No.	Chemical	M_r	VWV^a	SA^b	IP^c	$B \log P^d$	$-\Delta H^e$	VWV^f	$\log P^g$	$-\Delta H^h$	pK_a^i
199	2-Chlorobenzoic acid	156.568	124.325	159.776	9.931	1.991	5930	74.84	—	2.566	3.380
200	3-Chlorobenzoic acid	156.568	124.254	159.033	9.946	2.135	5870	74.84	—	2.704	3.960
201	4-Chlorobenzoic acid	156.568	124.250	159.022	10.023	2.143	5864	74.84	2.517	2.718	3.960
202	2,4-Dichlorobenzoic acid	191.013	137.827	175.267	10.066	2.538	6647	84.32	—	2.862	—
203	2,5-Dichlorobenzoic acid	191.013	137.947	176.071	9.847	2.495	6666	84.32	—	2.918	—
204	2,6-Dichlorobenzoic acid	191.013	137.710	174.850	9.971	2.398	6674	84.32	—	2.619	—
205	3,4-Dichlorobenzoic acid	191.013	137.721	173.886	9.989	2.655	6591	84.32	—	3.008	3.600
206	3,5-Dichlorobenzoic acid	191.013	137.913	175.409	10.010	2.607	6609	84.32	—	2.964	3.460
207	2-Bromobenzoic acid	201.019	132.145	166.909	9.952	2.300	6122	77.96	—	2.677	—
208	3-Bromobenzoic acid	201.019	132.23	166.492	9.969	2.491	6045	77.96	—	2.797	3.810
209	4-Bromobenzoic acid	136.150	132.209	166.737	9.824	2.488	6047	77.96	2.726	2.828	3.980
210	Phenylacetic acid	136.150	127.527	163.937	9.682	2.037	6010	79.55	1.94	2.428	4.300
211	2-Methylphenylacetic acid	150.177	143.730	182.111	9.476	2.413	7073	86.74	—	2.441	—
212	3-Methylphenylacetic acid	150.177	144.013	184.65	9.443	2.443	7188	86.74	—	2.486	4.325
213	4-Methylphenylacetic acid	150.177	144.143	185.405	9.383	2.470	7184	86.74	—	—	—
214	2-Ethylphenylacetic acid	164.204	160.255	201.187	9.436	2.776	7776	—	—	—	—
215	3-Ethylphenylacetic acid	164.204	160.509	204.385	9.361	2.846	8390	—	—	—	—
216	4-Ethylphenylacetic acid	164.204	160.492	204.073	9.310	2.838	7885	—	—	—	—
217	2-Chlorophenylacetic acid	170.595	140.569	177.038	9.643	2.564	6751	85.07	—	2.599	—
218	3-Chlorophenylacetic acid	170.595	140.759	178.739	9.622	2.568	7124	85.07	—	—	—
219	4-Chlorophenylacetic acid	170.595	140.631	178.664	9.518	2.495	7124	85.07	2.473	2.777	4.216
220	3,4-Dihydroxyphenylacetic acid	168.149	143.030	181.183	8.890	1.498	7667	—	0.98	—	4.387
221	2,5-Dihydroxyphenylacetic acid	168.149	143.039	182.608	8.810	1.360	7620	—	0.80	—	3.964
222	2-Hydroxyphenylacetic acid	152.149	134.959	171.653	9.195	1.726	6783	—	1.47	—	4.225
223	4-Hydroxyphenylacetic acid	152.149	135.192	173.179	9.106	1.807	6788	—	1.26	—	4.433
224	4-Hydroxy-3-methoxyphenylacetic acid	182.176	160.492	203.920	8.752	1.872	8539	—	1.18	—	4.394
225	3-Methoxyphenylacetic acid	166.176	152.512	194.122	9.111	2.111	7675	—	1.99	—	4.261
226	4-Methoxyphenylacetic acid	166.176	152.489	193.786	8.992	2.253	7709	—	1.95	—	4.394
227	3,4-Dihydroxymandelic acid	184.148	151.882	195.496	8.937	1.145	8638	—	0.144	—	3.493

No.	Compound										
228	Mandelic acid	152.149	135.783	173.022	9.991	1.752	6650	80.18	—	1.650	3.380
229	4-Hydroxy-3-methoxymandelic acid	198.175	168.774	213.274	8.784	1.551	9369	—	0.84	—	3.507
230	3-Methoxymandelic acid	182.176	160.809	203.411	9.088	1.892	8558	—	1.33	—	3.330
231	3,4-Dihydroxycinnamic acid	180.160	154.457	197.932	8.947	1.762	8310	—	1.15	—	4.545
232	4-Hydroxycinnamic acid	164.160	146.507	188.227	9.112	2.016	7416	—	1.46	—	4.630
233	*trans*-Cinnamic acid	148.161	138.435	177.359	9.472	2.333	6676	82.32	2.32	2.710	4.380
234	3-Methoxycinnamic acid	178.187	163.810	209.030	9.216	2.539	8326	—	2.37	—	4.307
235	4-Methyl(*trans*)cinnamic acid	162.188	155.116	198.854	9.259	2.751	7826	—	—	—	—
236	4-Hydroxy-3-methoxycinnamic acid	194.187	171.801	219.021	8.854	2.230	9159	—	1.51	—	4.558
237	4-Methylcinnamic acid	162.188	155.207	199.577	9.253	2.765	7851	93.47	—	2.789	4.492
238	4'-Methoxy-3-methoxycinnamic acid	208.213	189.048	239.781	8.755	2.616	10039	—	2.063	—	4.673
239	Phenylpropionic acid	150.177	144.183	185.755	9.520	2.454	6948	86.04	2.250	2.601	4.579
240	Phenylbutyric acid	164.204	160.924	207.898	9.434	2.890	7853	96.05	2.522	2.609	4.719
241	Phenylpentanoic acid	178.230	177.872	229.559	9.411	3.327	8727	—	—	—	—
242	3-Indoleacetic acid	161.160	139.965	176.493	8.825	0.864	6419	91.65	1.92	3.033	4.590
243	3-Indolepropionic acid	175.187	156.451	195.771	8.445	1.380	7416	114.11	—	3.281	—
244	3-Indolebutyric acid	189.213	175.527	219.961	8.431	1.604	8407	121.34	—	3.388	—
245	Hippuric acid	179.175	158.769	203.669	10.020	1.573	8184	96.15	—	1.518	—
246	2-Hydroxyhippuric acid	195.174	166.550	212.370	9.569	1.038	8802	—	1.55	—	3.506
247	Uric acid	168.112	123.569	159.147	9.165	-1.064	6557	—	—	—	—
248	2-Naphthoic acid	172.183	154.456	190.262	9.063	2.820	6766	—	—	—	—
249	3-Hydroxy-2-naphthoic acid	188.182	162.354	199.135	8.903	2.311	7575	—	—	—	—

[a] van der Waals volume ($Å^3$ mol^{-1}); [b] surface area ($Å^2$); [c] ionization potential (eV) and [d] partition coefficient are calculated by MOPAC-BlogP from CACheTM; [e] enthalpy are calculated by MOPAC from CACheTM; [f] van der Waals volume calculated by Bondi's method; [g] partition coefficient calculated by Rekker's method; [h] enthalpy measured by liquid chromatography; [i] dissociation constant.

The ionization potential calculated by MOPAC seemed to be capable of correcting the $\log k$ values of polychlorobenzenes when alkylbenzenes were used as standards, as shown in Figure 6.5 where the ionization potentials of alkyl and polychlorobenzenes are shown.

The relationship between the van der Waals volume (V) and the $\log k$ values of the alkylbenzenes is given by:

$$\log k = A \times V + B \qquad (6.9)$$

The constants A and B are related to the percentage concentration of acetonitrile (x), and their values are:

$$A = 1.1503 \times 10^{-2} + 4.9525 \times 10^{-5}x - 1.0825 \times 10^{-6}x^2 \qquad (r^2 = 1.000)$$
$$B = 1.305 - 3.2936 \times 10^{-2}x + 1.2900 \times 10^{-4}x^2 \qquad (r^2 = 1.000)$$

For hypothetical alkylbenzenes having the same van der Waals volume as alkanes, polycyclic aromatic hydrocarbons, and polychlorobenzenes, the calculated $\log k$ values were determined by means of Equation 6.9 with constants A and B obtained from the alkylbenzene results. The difference ($\Delta \log k$) between the measured and the calculated values of $\log k$ in the same eluent was determined.

The ionization potential energy, IP, of alkylbenzenes calculated by MOPAC was used as the standard, and it was almost constant for a series of alkyl compounds. The difference in the ionization potential, ΔIP, of other compounds from that of alkylbenzenes was calculated. The relationship between $\Delta \log k$ and ΔIP is given by:

$$\Delta \log k = C \Delta \text{IP} + D \qquad (6.10)$$

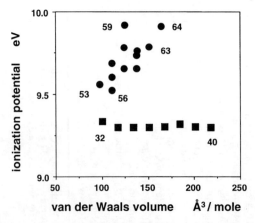

Figure 6.5 *Relationship of ionization potential to van der Waals volume. Compounds: ■, alkylbenzenes; and ●, polychlorobenzenes. Number beside symbols: see Table 6.4.*

where C and D are constants in the same eluent. The relationship between C, D, and the acetonitrile concentration (x) was obtained from

$$C = 0.266 + 2.826 \times 10^{-3}x - 3.640 \times 10^{-5}x^2 \qquad (r^2 = 0.999) \qquad (6.11)$$

$$D = -0.113 + 2.177 \times 10^{-3}x - 8.798 \times 10^{-6}x^2 \qquad (r^2 = 0.989) \qquad (6.12)$$

The $\log k$ of alkanes, polycyclic aromatic hydrocarbons, and polychlorobenzenes was therefore estimated from Equations 6.9 and 6.10 in 60–90% aqueous acetonitrile. The calculated equation is:

$$\log k_{calc} = Ax + B + C\Delta IP + D \qquad (6.13)$$

where A, B, C, and D are constants. The correlation coefficient between the measured and calculated retention factors was 0.922 ($n = 27$), 0.946 ($n = 32$), 0.966 ($n = 35$), and 0.971 ($n = 37$) in 60, 70, 80, and 90% aqueous acetonitrile, respectively.

This method was applied to the prediction of the retention factor of phenols. First, this method requires a relationship between $\log k$ and the van der Waals volume of a homologous series of alkyl compounds such as alkylbenzenes or alkylphenones. When alkylbenzenes are used as the standard, the value of B in Equation 6.8 should be altered to give a parallel relationship at a suitable position (E) on the y-axis for a different group of compounds. When B is moved to E for phenols, the $\log k$ values of *para*-alkylated phenols (y) are simply predicted from their van der Waals volume by the following equation:

$$\log k_{calc} = Ax + E \qquad (6.14)$$

For other substituted phenols, the difference in their ionization potential from that of phenol was calculated, and then their $\Delta \log k$ was obtained from Equation 6.10. The $\log k$ of substituted phenols was then estimated from Equations 6.9 and 6.13.

The retention time of phenols was predicted in 70 and 60% acidic aqueous acetonitrile on an ODS silica gel column. The constants A, B, C, and D were obtained from the above equations. The result in 60% aqueous acetonitrile is shown in Figure 6.6. The correlation coefficients between the measured and predicted retention factors of substituted phenols in 60 and 70% acidic aqueous acetonitrile were 0.974 ($n = 36$) and 0.967 ($n = 36$), respectively. In this system, the values of the slopes, which indicate the relationship between the measured and predicted retention factors, were 0.81 and 0.94 in 60 and 70% acidic aqueous acetonitrile, respectively.[32]

This system of using the van der Waals volume and ionization potential is much simpler than previous systems. It requires only one homologous series of alkyl compounds and one standard such as phenol for substituted phenols. The precision could be improved for polar compounds if *para*-alkylphenols and *para*-alkylbenzoic acids are used; however, only a few such compounds are

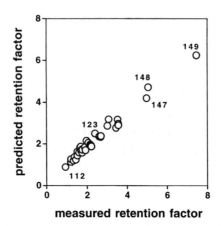

Figure 6.6 *Comparison of measured and predicted retention factors of substitute phenols. Column, 5 µm ODS-bonded silica gel, 15 cm × 4.6 mm i.d.; eluent, 60% aqueous acetonitrile containing 0.05 M phosphoric acid. Number beside symbols: see Table 6.4.*

available at present. This system seems to be practical for estimating the approximate retention time.

3 Enthalpy as a Thermodynamic Property in Retention Studies

The chromatographic retention parameters used in correlation studies are normally assumed to be proportional to the free energy change.

Adsorption is accompanied by a loss of surface free energy. Free energy change is related to enthalpy, ΔH, and entropy, ΔS, changes by the Gibbs equation:

$$\Delta G = \Delta H - T\Delta S \qquad (6.15)$$

The retention factor is related to the thermodynamic equilibrium constant K for solute binding by $k = \beta K$, where β is the phase ratio of the column. The free-energy change for the chromatographic process is expressed by

$$\Delta G = -RT \ln K = -RT \ln(k/\beta) \qquad (6.16)$$

where R is a gas constant ($8.314 \, \text{J K}^{-1} \, \text{mol}^{-1}$) (1 cal is 4.184 J). From Equations 5.38 and 5.39:

$$\ln k = -(\Delta H/RT) + (\Delta S/R) + \ln \beta \qquad (6.17)$$

A van't Hoff plot of $\ln k$ against $1/T$ yields a straight line if the stationary phase is not deformed. The relation $\ln k$ and $-\Delta H/RT$ is used to study the temperature-dependent selectivity of the stationary phase.[33-35]

Measurement of Enthalpy by Liquid Chromatography

Enthalpy can be measured by liquid chromatography where enthalpy is a slope of the relationship between $\ln k$ and the inverse value of the absolute temperature. A schematic diagram is shown in Figure 6.7. The slope depends upon the solutes being retained by the same liquid chromatographic mechanism. An example is given in Table 6.4. The results, measured on an octadecyl-bonded vinyl alcohol copolymer gel, did not show a simple linear relationship. This is due to a conformation change of the octadecyl-bonded vinyl alcohol copolymer gel stationary phase material, which has a phase transition point at about 33 °C.

Enthalpy Related to Retention

The higher the enthalpy change (ΔH), the longer the retention, and $\log k$ shows a linear correlation to enthalpy.[33] As seen in Figure 6.8, a number of groups of compounds exhibit such a linear relationship between enthalpy and $\log k$, although it is not perfect. The ΔH values were independent of their van der Waals volumes, as seen in Figure 6.9.

The $\Delta \Delta H$ value of a methylene unit of alkanols is fairly constant.[36] However, the $\Delta \Delta H$ value of a methylene unit of alkylbenzenes was not constant and the larger molecules had a higher $\Delta \Delta H$ value than the average $\Delta \Delta H$ value of a methylene unit. This larger value may be due to dimerization between solutes as described schematically,[37] and/or to a strong direct interaction between alkylbenzenes and the octadecyl group of the stationary phase material, because alkylbenzenes with a longer alkyl group must lose their aromaticity.

The isomeric effect of the $\Delta \Delta H$ value of substituted chlorobenzenes was very small. This means that the molecular size of the isomers is almost the same; however, the actual size is not the same under these chromatographic conditions.[38]

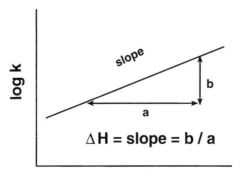

Figure 6.7   *Measurement of enthalpy using chromatography for the relationship between absolute temperature and retention factor.*

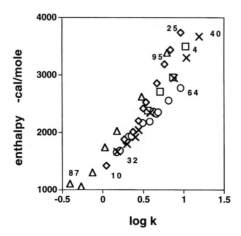

Figure 6.8   *Relationship of enthalpy to log k values. Column, ODS silica, ERC-ODS, 15 cm × 6.0 mm i.d.; eluent, 80% aqueous acetonitrile at 30 °C. Numbers beside symbols: see Table 6.4. ◇, Polycyclic aromatic hydrocarbons; ×, alkylbenzenes; ○, halogenated benzenes; △, alkanols; and □, alkanes.*

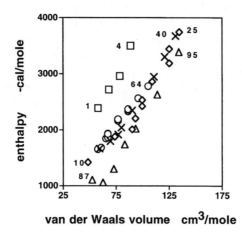

Figure 6.9   *Relationship of enthalpy to Bondi's van der Waals volumes. Column, ODS silica, ERC-ODS, 15 cm × 4.6 mm i.d.; eluent, 80% aqueous acetonitrile. Numbers beside symbols: see Table 6.4. ◇, Polycyclic aromatic hydrocarbons; ×, alkylbenzenes; ○, halogenated benzenes; △, alkanols; and □, alkanes.*

The enthalpy of methylphenols was about 2.0 kcal mol^{-1}, and that of chlorophenols varied from 2.0 to 2.4 kcal mol^{-1} in the case of pentachlorophenol, indicating that the retention difference depended not upon the size but on the π-electron density.[39] A similar result was obtained for alkylated and halogenated aromatic acids, whose enthalpies were nearly equal, but whose retention factors were different.[40] The ΔH values may depend on the type of stationary phase used and the water content of the eluent.[41]

The highest enthalpy was found on non-endcapped high carbon octadecyl-bonded silica gels, and was slightly higher on high-carbon endcapped gels than on low-carbon endcapped gels. There was a selectivity difference in the retention between alkanols and other hydrophobic compounds that depended on the end-capping treatment applied.[42] The order of enthalpies of phenols showed no direct relationship to the retention factor or hydrophobicity of octadecyl-bonded silica gels with different carbon loadings.[43] Further, enthalpy was not directly related to the $\log k$ values nor to the alkyl chain length of bonded-silica gels, although the retention behaviour of individual groups of compounds such as polycyclic aromatic hydrocarbons, and especially the retention of smaller-size polar compounds, appeared to indicate this.[44]

4 Conclusion

The qualitative analysis of retention behaviour in liquid chromatography has now become possible. Quantitative retention–prediction is, however, still difficult; the prediction of retention time and the optimization of separation conditions based on physicochemical properties have not yet been completely successful. One reason is the lack of an ideal stationary phase material. The stationary phase material has to be stable as part of an instrument, and this is very difficult to achieve in normal-phase liquid chromatography because the moisture in organic solvents ages the silica gel.

Several improved stationary phase materials have been synthesized for reversed-phase liquid chromatography. One material is vinyl alcohol copolymer gel. This stationary phase is quite polar and chemically very stable; however, it demonstrated a strong retention capacity for polycyclic aromatic hydrocarbons.[45–49] Although stable octadecyl- and octyl-bonded silica gels have been synthesized from pure silica gel[50,51] and are now commercially available, such an optimization system has not yet been built. Further experiments are required to elucidate the retention mechanism, and to systematize it within the context of instrumentation.

5 References

1. A. Leo, C. Hansch, and D. Elkins, *Chem. Rev.*, 1971, **71**, 525.
2. R.F. Rekker, 'The Hydrophobic Fragmental Constant', Elsevier Scientific, Amsterdam, 1977.
3. W.J. Haggerty and E.A. Murrill, *Res. Dev.*, 1974, **25**, 30.
4. R.M. Calson, *J. Chromatogr.*, 1975, **107**, 219.
5. J.M. McCall, *J. Med. Chem.*, 1975, **18**, 549.
6. M.S. Mirrless, S.T. Moulton, C.T. Murphy, and P.J. Taylor, *J. Med. Chem.*, 1976, **19**, 615.
7. K. Miyake and H. Terada, *J. Chromatogr.*, 1978, **157**, 386.
8. M.J. O'Hare and E.C. Nice, *J. Chromatogr.*, 1979, **171**, 209.
9. J. Grunenberg and R. Herges, *J. Chem. Inf. Comput. Sci.*, 1995, **35**, 905.
10. C. Yamagami, M. Yokora, and N. Takao, *J. Chromatogr.*, 1994, **662**, 49.
11. A. Kakoulidou and R.F. Rekker, *J. Chromatogr.*, 1984, **295**, 341.
12. M.J.M. Wells, C.R. Clark, and R.M. Patterson, *J. Chromatogr. Sci.*, 1981, **19**, 573.

13. M.D. Amboise and T. Hanai, *J. Liq. Chromatogr.*, 1982, **5**, 229.
14. K. Calmmer, L.-E. Edholm, and B.E.F. Smith, *J. Chromatogr.*, 1977, **136**, 45.
15. J.F. Schabron, R.J. Hurtubise, and H.F. Silver, *Anal. Chem.*, 1980, **50**, 1911.
16. N. Nomura, *Toyama Univ. Kyoikugakubu-Kiyo*, 1979, **27**, 1.
17. B.L. Karger, J.R. Gant, A. Harkopt, and P.H. Weiner, *J. Chromatogr.*, 1976, **128**, 65.
18. T. Hanai, *Chromatographia*, 1979, **12**, 77.
19. T. Hanai, K.C. Tran, and J. Hubert, *HRC & CC*, 1981, **4**, 454.
20. T. Hanai, S. Hara, S. Mori, and T. Hanai (eds), 'Chromatography, the Separation System', Maruzen, Tokyo, 1981, p. 138 (in Japanese).
21. T. Hanai and J. Hubert, *HRC & CC*, 1983, **6**, 20.
22. T. Hanai and J. Hubert, *J. Liq. Chromatogr.*, 1985, **8**, 2463.
23. T. Hanai and J. Hubert, *J. Chromatogr.*, 1982, **239**, 527.
24. T. Hanai and J. Hubert, *HRC & CC*, 1981, **4**, 500.
25. T. Hanai, K.C. Tran, and J. Hubert, *J. Chromatogr.*, 1982, **239**, 385.
26. T. Hanai and J. Hubert, *HRC & CC*, 1984, **7**, 524.
27. D.D. Perrin, B. Dempsey, and E.P. Serjeant, 'pK_a Prediction for Organic Acids and Bases', Chapman & Hall, London, 1981.
28. A. Bondi, *J. Phys. Chem.*, 1964, **68**, 441.
29. T. Hanai, *J. Chromatogr.*, 1985, **332**, 189.
30. T. Hanai, H. Hatano, N. Nimura, and T. Kinoshita, *J. Liq. Chromatogr.*, 1993, **16**, 1453.
31. J. Yamaguchi, T. Hanai, and H. Cai, *J. Chromatogr.*, 1988, **441**, 183.
32. T. Hanai, H. Hatano, N. Nimura, and T. Kinoshita, *Analyst*, 1994, **119**, 1167.
33. W. Melander, D.E. Campbell, and C. Horvath, *J. Chromatogr.*, 1978, **158**, 215.
34. T. Paryjczak, 'Gas Chromatography in Adsorption and Catalysis', Ellis Horwood Limited, New York, 1986, p. 231.
35. R. Kaliszan, 'Quantitative Structure–Chromatographic Retention Relationships', Wiley, New York, 1987.
36. T. Hanai and J. Hubert, *J. Chromatogr.*, 1984, **291**, 81.
37. A. Ben-Neim, 'Hydrophobic Interaction', Plenum Press, New York, 1980.
38. T. Hanai, A. Jukurogi, and J. Hubert, *Chromatographia*, 1984, **19**, 266.
39. Y. Arai, M. Hirukawa, and T. Hanai, *J. Chromatogr.*, 1987, **384**, 279.
40. Y. Arai, J. Yamaguchi, and T. Hanai, *J. Chromatogr.*, 1987, **400**, 21.
41. H. Colin, J.C. Diez-Mesa, G. Guiochon, T. Czaikowska, and I. Miedziak, *J. Chromatogr.*, 1978, **167**, 41.
42. J. Yamaguchi, T. Hanai, and H. Cai, *J. Chromatogr.*, 1988, **441**, 183.
43. J. Yamaguchi and T. Hanai, *J. Chromatogr. Sci.*, 1989, **27**, 710.
44. J. Yamaguchi and T. Hanai, *Chromatographia*, 1989, **27**, 371.
45. T. Hanai, Y. Arai, M. Hirukawa, K. Noguchi, and Y. Yanagihara, *J. Chromatogr.*, 1985, **349**, 323.
46. Y. Arai, M. Hirukawa, and T. Hanai, *Nihon Kagaku Kaishi*, 1986, 969.
47. Y. Arai, M. Hirukawa, and T. Hanai, *J. Liq. Chromatogr.*, 1987, **10**, 635.
48. M. Hirukawa, Y. Arai, and T. Hanai, *J. Chromatogr.*, 1987, **395**, 481.
49. Y. Arai, M. Hirukawa, and T. Hanai, *J. Chromatogr.*, 1987, **400**, 27.
50. T. Hanai, M. Ohhira, and T. Tamura, *LC & GC*, 1988, **6**, 922.
51. M. Ohhira, F. Ohmura, and T. Hanai, *J. Liq. Chromatogr.*, 1989, **12**, 1065.

Subject Index